Geology of the Wallowa Mountains of Oregon: Part I

*Compiled by the staff of
the Oregon Department of Geology and Mineral Industries*

with an introduction by Kerby Jackson

*This work contains material that was originally published
in 1938 by DOGAMI.*

*This publication was created and published for the public benefit,
utilizing public funding and is within the Public Domain.*

*This edition is reprinted for educational purposes
and in accordance with all applicable Federal Laws.*

Introduction Copyright 2014 by Kerby Jackson

Introduction

It has been over seventy five years since the Oregon Department of Geology and Mineral Industries released it's important publication "Geology of Part of the Wallowa Mountains of Oregon". First released in 1938, this important volume has now been out of print for years and has been unavailable to the mining community since those days, with the exception of expensive original collector's copies and poorly produced digital editions.

It has often been said that "*gold is where you find it*", but even beginning prospectors understand that their chances for finding something of value in the earth or in the streams of the Golden West are dramatically increased by going back to those places where gold and other minerals were once mined by our forerunners. Despite this, much of the contemporary information on local mining history that is currently available is mostly a result of mere local folklore and persistent rumors of major strikes, the details and facts of which, have long been distorted. Long gone are the old timers and with them, the days of first hand knowledge of the mines of the area and how they operated. Also long gone are most of their notes, their assay reports, their mine maps and personal scrapbooks, along with most of the surveys and reports that were performed for them by private and government geologists. Even published books such as this one are often retired to the local landfill or backyard burn pile by the descendents of those old timers and disappear at an alarming rate. Despite the fact that we live in the so-called "Information Age" where information is supposedly only the push of a button on a keyboard away, true insight into mining properties remains illusive and hard to come by, even to those of us who seek out this sort of information as if our lives depend upon it. Without this type of information readily available to the average independent miner, there is little hope that our metal mining industry will ever recover.

This important volume and others like it, are being presented in their entirety again, in the hope that the average prospector will no longer stumble through the overgrown hills and the tailing strewn creeks without being well informed enough to have a chance to succeed at his ventures.

Kerby Jackson
Josephine County, Oregon
July 2014

Scene in the Wallowas
Courtesy: United States Forest Service

STATE OF OREGON
DEPARTMENT OF GEOLOGY AND
MINERAL INDUSTRIES

Governing Board
W. H. Strayer, Albert Burch, E. B. MacNaughton

Earl K. Nixon, Director

BULLETIN NUMBER 3

The Geology of Part of the Wallowa Mountains

By C. P. ROSS of the

UNITED STATES GEOLOGICAL SURVEY

This publication may be obtained by addressing the State Department of Geology and Mineral Industries, 704 Lewis Building, Portland, Oregon.

PRICE, 50c

PREFACE

The field work on which this report is based was done in 1921 by Mr. C. P. Ross of the United States Geological Survey under a cooperative arrangement whereby the Oregon Bureau of Mines and Geology shared the expense with the Federal Survey. Owing to the discontinuance of the Oregon Bureau of Mines and Geology, funds were not available for publishing the report.

On account of renewed interest, especially in the search for tungsten and molybdenum, and also for the reason that little has been published on the general geology of this Wallowa area, it was deemed expedient to publish the pertinent data at this time. Dr. W. C. Mendenhall, Director of the United States Geological Survey, kindly consented to have the portion of the report covering the general geology segregated and prepared for publication. This bulletin is the result.

On account of recent underground development of various old mines and the finding of some new ones, the description of the mineral resources of the area should be brought up to date. This work is contemplated for the 1938 field season.

The State Department of Geology and Mineral Industries wishes to express appreciation of the cooperation of the United States Geological Survey in its preparation of this bulletin.

EARL K. NIXON, Director.

704 Lewis Building, Portland, Oregon.
January, 1938.

CONTENTS

	PAGE
Abstract	7
Scope	7
Stratigraphy	7
Structure	7
Geologic History	7
Introduction	9
Purpose and Scope of the Report	9
The Map	9
Surface Features	10
Climate	13
Industries	16
Stratigraphy and Petrology	17
Summary Note	17
Paleozoic Stratified Rocks	20
Black Slate	20
Clover Creek Greenstone	21
Carboniferous (?) Sedimentary Rocks	26
Epidote-garnet Rock	31
Mesozoic Stratified Rocks	32
Martin Bridge Formation	32
Triassic (?) Volcanic Rocks	36
Younger Mesozoic Sedimentary Rocks	38
Hornfels	40
Mesozoic and Older Intrusive Rocks	42
Metamorphosed Dike Rocks	42
Amphizolite	43
Diorite-gabbro Complex	44
Albite Granite	45
Quartz Diorite and Granodiorite	47
Aplite and Related Rocks	50
Cenozoic Stratified Rocks	52
Columbia River Basalt	52
Tuff	53
Clastic Rocks in Eagle Valley	54
Tertiary Dike Rocks	56
Unconsolidated Sediments	57
Structure	61
Pre-tertiary Earth Movements	61
Structure in the Tertiary Rocks	68
Geologic History	69
Pre-carboniferous Time	69
Carboniferous Time	70
Mesozoic Time	71
Tertiary Time	73
Quaternary Time	74

ILLUSTRATIONS

Plate		Page
1.	Geologic map and structure section of a portion of the Wallowa Mountains, Oregon	In Pocket
2.	Two Granites; near Cornucopia, Oregon. Also known as The Granites, Twin Granites, and Cornucopia Mountain. It is composed of granodiorite. Cornucopia in the foreground. The larger mines of the district are visible in the picture. Photo by Lawrence Panter	14
3. A.	Crater Lake. This lake fills a depression made by glacial scour in granodiorite. View looking northeast from Truax Mountain. Photo by Lawrence Panter	18
B.	Two Granites and Simmons Mountain from the plateau to the east. Note contrast in topography. Photo by Jaques Heupgen	18
4. A.	East side of the valley of upper Clear Creek. Effects of glacial erosion of rock of the diorite-gabbro complex in the plateau	19
B.	Park at Fish Lake Ranger Station. A typical park in the "Plateau Area"	19
5. A.	The upland surface in "Powder River Valley" near Pleasant Ridge School	24
B.	View west in Pine Valley from near Sunset. Note terrace in foreground and sharp contrast in the appearance of the valley and its enclosing hills	24
6. A.	The irregular contact of the stock on the West Fork of Pine Creek. The granitic rock is on the left of the picture. The strata on the right belongs to the younger Mesozoic sedimentary rocks	48
B.	Bonanza Basin between the Last Chance and Queen of the West mines; view taken from the latter shows inclusion of hornfels in granodiorite, basalt dike is visible in the distance	48

FIGURES

Figure		Page
1.	Outline map of Oregon showing location of the "Wallowa Region"	11
2.	Outline map showing major tectonic lines in the "Wallowa Region"	62

[5]

THE GEOLOGY OF PART OF THE WALLOWA MOUNTAINS, OREGON

by

CLYDE P. ROSS

Abstract

Scope.—This report gives the results of a geologic reconnaissance made in 1921 of about 460 square miles in the southern part of the Wallowa Mountains. The region covered includes the Cornucopia gold mining district and other smaller mining districts, but the ore deposits are not discussed in the present paper. The rocks of the region have been divided on the map into 19 lithologic and stratigraphic units, and the structure and history of these units is discussed.

Stratigraphy.—The rocks range in age from Paleozoic (Carboniferous or possibly older) to Recent. The Paleozoic rocks in the high mountains consist of a great series of lava flows and pyroclastics with sedimentary beds of probable shallow water origin overlying them. East of the high mountains is a large area of sedimentary rocks which may be in part older than the volcanics. Upper Triassic limestones with some shale and minor amounts of associated volcanic rocks rest unconformably on the Paleozoic strata. Above these are younger Mesozoic sedimentary rocks. These rocks, especially the older ones, are metamorphosed in varying degrees. They have been intruded by quartz diorite and granodiorite, albite granite, and by a diorite-gabbro complex. These different rocks, all of which may be related to each other, are clearly of Mesozoic age and probably mainly intruded during or just before the Cretaceous period. There is also an intrusion of some calcic rock now altered to amphibolite, which may be older. The Columbia River basalt overlies the Mesozoic and Paleozoic rocks and covers a large part of the region outside of the mountains proper. West of Sparta tuff, and in Eagle Valley beds of sand, gravel, and diatomaceous earth are intercalated in the lava. The alluvial filling of Pine and Eagle Valleys and the stream and glacial deposits are of even later date.

Structure.—The dominant structural feature of the pre-Tertiary rocks is folding, which in the mountains has been so intense as to result in overturning. The average strike of the folds is about N. 26° E. in the mountains, swinging more to the east farther south. The strike of the larger faults is about N. 30° W. The Tertiary rocks have been warped, but few definite folds have been produced. Faults in the Tertiary rocks almost surround Pine Valley, and small ones were noted in various other places.

Geologic History.—None of the exposed rocks are known to be older than Carboniferous. During much of the Paleozoic era the region was probably part of a land mass. In the Carboniferous period it was sub-

---

[1] Published with the permission of the Director, U. S. Geological Survey.

merged, but there were oscillations, some of which may have brought it temporarily above the sea. At this time there was a period of great volcanic activity. The Carboniferous ended with an epoch of strong folding. After an erosion interval marine Mesozoic deposits accompanied in places by lava and tuff were formed. This was followed near the end of the Mesozoic by folding, batholithic intrusion and uplift. The Wallowa Mountains and other present day ranges were born at this time. Later, after a sequence of erosional events not here discussed, successive flows of basaltic lava overlapped the flanks of the mountains and partially buried them. Some of these flows were erupted through fissures within the mountains, now occupied by dikes. The subsequent history has been one of erosion and glaciation, with comparatively minor earth movements.

INTRODUCTION

Purpose and Scope of the Report

As shown in figure 1, the area covered by this report lies in the southeastern part of the Wallowa Mountains in Baker and Wallowa Counties in northeastern Oregon. It comprises the northern part of Pine quadrangle and the southern part of the as yet unsurveyed Wallowa Lake quadrangle of the Topographic Atlas of the United States. The greater portion lies in the Minam Division of the Whitman National Forest. In this report it will, for convenience, be referred to as the "Wallowa Region".

Because the original plans could not be carried out either as to area to be covered or as to details to be secured, the present bulletin may be regarded as a preliminary report on a region of such complex geology as to warrant more intensive study than it has yet been possible to give it.

The field study of the areal geology occupied the time from the middle of July to the end of September, 1921, and was done under the auspices of the Oregon Bureau of Mines and Geology. The writer was assisted in the field by Jacques Heupgen. Travel was carried on by means of a pack train owned and operated by Aaron A. Densley, whose efficient services aided in the prosecution of the work. Uniform kindness, hospitality, and assistance were extended by everyone encountered during the field season. The report has been prepared in the office of the United States Geological Survey. The microscopic work on the granular intrusive rocks was kindly checked by C. S. Ross, of the United States Geological Survey.

The Map

The topographic map of Pine quadrangle prepared by the United States Geological Survey was used as a base for the geologic mapping south of the 45th parallel of latitude. North of the 45th parallel the reconnaissance topographic map of the Forest Service was the only large map available. Where this map was too inaccurate to show the geology satisfactorily it was revised by field sketching. Such revisions were made on

Kettle Creek, West Basin,[2] and the head of Imnaha River. The character of the country and of the rocks is such that it was not possible in the time available or on the base map used to search out and map all small outliers of the various formations. Several patches of Columbia River basalt and of some of the other formations have been omitted from the map because they are too small to show without exaggeration. Where the positions of geologic boundaries have been determined with less than the average accuracy they are shown with dashed lines. Inaccurate location of a boundary may be due to soil, talus, or vegetation cover, to lack of sharpness in the contact itself, or to difficulty of access. Some inaccuracies arising from errors in the base map are also present.

Surface Features

The Wallowa region may, for purposes of description, be divided into four topographic units. These are, first, the rugged mountain area; second, the adjacent dissected plateau; third, the semi-arid country along Powder River; and, fourth, the alluvium-filled Pine and Eagle valleys.

The mountains are intricately sculptured and possess local relief of 2,000 to 4,000 feet. The highest point is Eagle Cap, 9,675 feet above the sea and numerous peaks reach altitudes of over 8,000 feet. The difference in altitude between the mouth of Powder River and the top of Eagle Cap is about 7,800 feet. This large difference within so short a distance has been a potent factor in dissection. In consequence of this and of the local over-steepening produced by the extinct glaciers, cliffs and pinnacles are abundant in the mountains. The character of the dissection varies with the character of the bedrock. For example, in such a mass of granodiorite and quartz diorite as that illustrated in Plate 2, slopes are steep but cliffs and pinnacles are less common than in many areas underlain by stratified rocks. Most of the country above an altitude of 8,000 feet and, locally, several hundred feet lower is nearly or quite free from soil, a fact which is also shown by Plate 2. Below these rocky areas most of the mountain slopes are

---

[2] Corrections in the topography of West Basin as represented in Plate 1 are based mainly on a map by C. G. Dobson, formerly consulting engineer for the Gold Reef Mining Co.

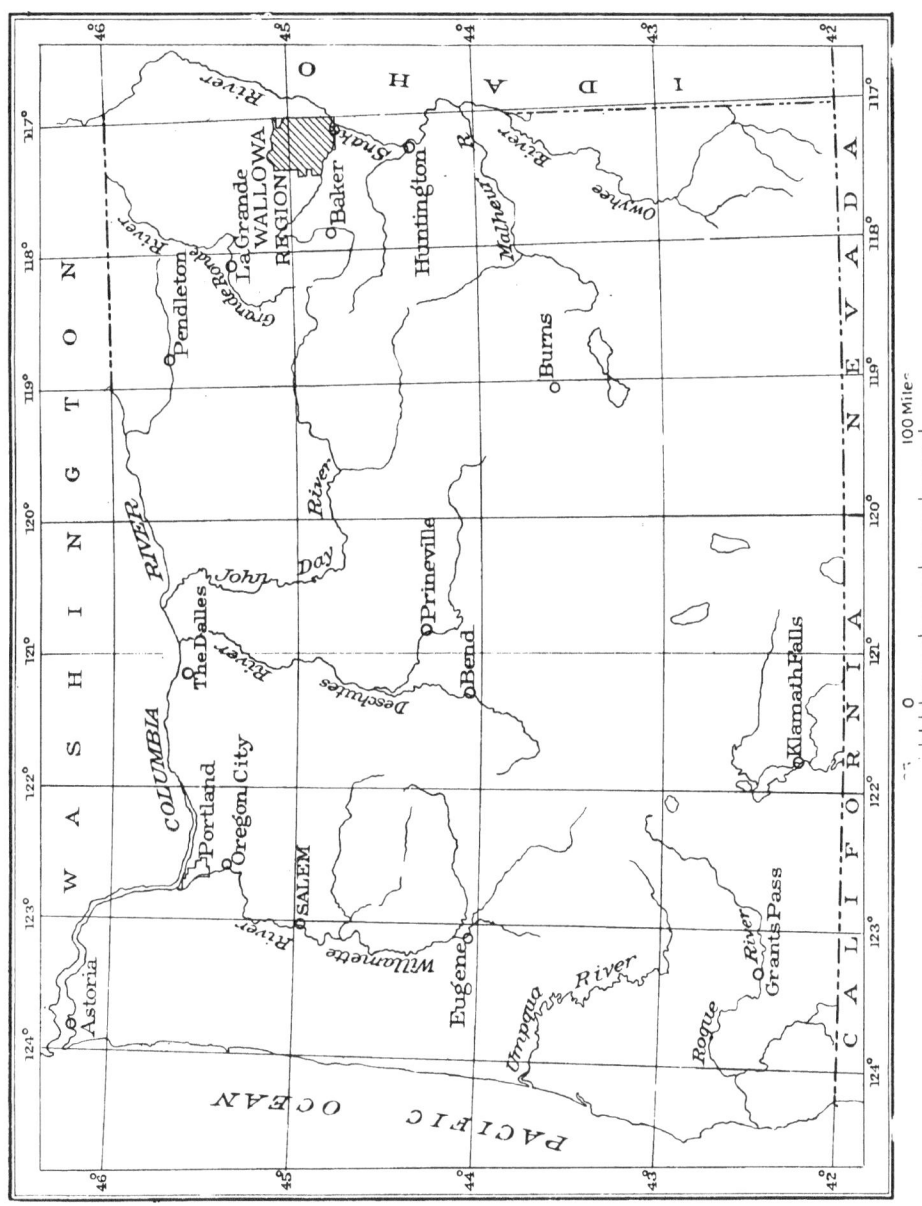

Figure 1. Outline map of Oregon showing location of the "Wallowa region".

somewhat sparsely covered with a forest of yellow pine, fir, tamarack and other conifers, with considerable brush along the streams; especially in the southern part. The stream valleys within the mountains are steep-sided and generally U-shaped in cross section. Most of those that extend to high altitudes end in benched, rock-floored basins, many of which contain lakes like that shown in Plate 3-a. The lake shown in this photograph has been tapped by a tunnel near its up-stream end for the purpose of obtaining water for irrigation. In contrast to those at lower levels, most of the streams within the mountains are perennial.

The mountains are bordered on the east and south by a dissected plateau. The sharp contrast between the rolling interstream areas here and the steep slopes in the mountains is illustrated in Plate 3-B. Such broad upland surfaces are conspicuous above an altitude of 6,500 feet. Below this, erosion has proceeded far enough to destroy much of the old plateau leaving steep-sided but comparatively rounded ridges between the stream valleys. Even at higher altitudes dissection has been marked. Glaciation has left widespread evidence of its work. The bare rock surfaces, locally grooved and polished, that are conspicuous on the rolling uplands result from glaciation. Part of such a surface is shown in Plate 4-A. The vegetation in the plateau area is similar to that on the lower slopes of the mountains. The trees are generally smaller and the brush along streams is somewhat less dense. The upper reaches of the streams contain numerous open parks bordered by trees, like the one shown in Plate 4-B. The nearly flat surfaces of most of these parks are crossed by small meandering streamways in which water flows for much of the year. On account of the luxuriant growth of grass on their imperfectly drained surfaces, the parks add much to the value of the region for stock grazing. Some are so poorly drained that shallow, generally mud-floored lakes cover parts of them. Although these parks lie near the heads of present streamways they are not at grade with the modern streams, which commonly descend from them in a series of cascades. The parks are, at least in part, among the results of glaciation.

Beyond the dissected, partially forested plateau are the semiarid hills that border Powder River. For much of this part of the region, the range in altitude is 3,000 to 4,000 feet. Many of the hills are flat-topped and there are flat areas between them at different altitudes. Powder River, which forms the southern boundary of the area mapped, has cut a trench about 1,000 feet deep across the semi-arid country. It is much the largest stream included in the area shown on Plate 1, but in part because of the demands upon it for water for irrigation, has only a moderate discharge. Few of the numerous sharply incised stream valleys, tributary to Powder River in this region, are perennial beyond the borders of the dissected plateau above described. In most, water flows at the surface only in direct response to rains. Even Eagle Creek, the largest of the tributaries, is reported to be dry in its lower portion at times.

Eagle and Pine Valleys are small but distinct topographic units. Both are alluvium-filled structural basins in the Columbia River basalt, named for the streams which drain them. Both valleys have fairly flat floors, but both show terraces. In Eagle Valley there are three or more, doubtless connected in some way with changes in the level of Powder River. In Pine Valley only remnants of one terrace were observed. This terrace, which is in the northeast part of the valley, is shown in Plate 5-B. The upper terraced part of the alluvium in both valleys is sufficiently free from gravel to permit crops to be grown profitably. The green fields and prosperous settlements of these two valleys constitute garden spots whose quiet beauty is especially striking in contrast to the aridity of the surrounding basalt hills. Pine, Clear, and East Pine Creeks in Pine Valley flow almost parallel to each other for about six miles and unite near the lower end of the valley. The course of Dry Creek within the valley also trends southeastward.

Climate

The climate in this region varies with altitude and topography. In general the greater the altitude, the greater is the precipitation and the lower the average temperature. The prevailing wind is westerly, and localities so situated as to be protected from it have less than the average precipitation for their altitude.

At Cornucopia, on the edge of the mountain area, the average annual precipitation is about 45 inches. In exposed parts

Plate 2. Two Granites, near Cornucopia, Oregon. Also known as The Granites, Twin Granites, and Cornucopia Mountain. It is composed of granodiorite. Cornucopia in the foreground. The larger mines of the district are visible in the picture. (1) Union Companion mine. (2) Red Jacket mine. (3) Last Chance mine. (4) Queen of the West mine. (5) Mayflower mill.

(Photo by Lawrence Panter)

of the mountains the annual precipitation (mainly snow) is probably much more than this.

CLIMATOLOGICAL DATA FOR SPARTA, OREGON

| Year | Temperature ||||| Total annual precipitation (inches) | Total annual snow (inches) |
------	Annual mean degrees F.	Highest degrees F.	Date	Lowest degrees F.	Date		
1918	47.3	94	June 22	—10	Jan. 30	16.09	84.5
1919	45.2	100	July 16	—19	Dec. 13	17.99	133.5
1920	44.5	96	Aug. 15	3	Jan. 8	19.92	82.0
1921	46.0	92	Aug. 1	1	Feb. 16	13.88	93.0

The mildest and driest climate in the region is that of Eagle Valley. It is noticeably warmer in the summer here than in the mountains, although seldom oppressively hot. The valley is to a considerable extent protected from the winds which sweep across the surrounding uplands. Less snow is reported to fall here than elsewhere in the region visited. The following table, compiled from the records of the Weather Bureau, summarizes the data regarding the climate of Richland in the center of Eagle Valley. More recent records than those given in the two tables here presented are too incomplete to be of value.

CLIMATOLOGICAL DATA FOR RICHLAND, OREGON

| Year | Temperature ||||| Total annual precipitation (inches) | Total annual snow (inches) |
------	Annual mean degrees F.	Highest degrees F.	Date	Lowest degrees F.	Date		
1918	51.0	105	June 20	0	June 30	9.07	20.0
1919	49.1	106	July 9	—22	Dec. 13	9.85	17.0
1920	49.6	104	July 27	— 3	Jan. 11	11.17	....
1921	......	99	July 23	....	....	....	....

The first snow of the winter may be expected by the middle of September, but it is rare to have notable or lasting amounts of snow before October or November. In winter the temperature frequently drops below zero, but much colder weather is rare. June, July, and August usually have very little rain, and most of the storms which do occur take place at night or in the early morning. In summer it is warm, but rarely oppressively hot. The nights are always sufficiently cool to be comfortable.

In the lower plateau country, especially on the south side of the mountains, the average annual precipitation is about 18 inches, as indicated by the Weather Bureau records at Sparta. Most of the precipitation takes place in early winter, but there is also considerable rainy weather in spring. In the fall and winter months there is frequently enough wind to be disagreeable, but seldom enough to be a serious drawback.

Pine Valley is similar topographically to Eagle Valley, but has colder winters and more snow, probably because its average altitude is greater. Climatological records for Pine Valley are not available so that exact comparisons cannot be made.

Industries

The three principal industries of the Wallowa region are mining, stock raising, and farming. Lumbering sufficient to supply the local demand is carried on but it is not an important industry. Mining formerly outranked the others in importance, but in recent years has been declining although there have been periods of revival, one of which began in 1935. Grazing of sheep and cattle is the only important industry in the mountainous portions of the region. Most of the people have their homes in the lower country. Here they engage in farming incidental to stock raising.

The three farming localities are Pine Valley, Eagle Valley, and the Sparta district. Except for the residents of the mining town of Cornucopia and a few scattered prospectors, the population of the region is concentrated here during the winter. In the summer stockmen, prospectors, forest rangers, and pleasure seekers from these and more distant communities spread over the mountains.

The census of 1920 enumerated in Cornucopia 242 people, in Pine Valley 1,491 people, in Eagle Valley 1,263 people, and in Sparta 249 people, a total of 3,252, which represents nearly all the inhabitants of the region here considered. In 1930 the total population was 2,639, the principal loss being in Cornucopia precinct, whose population was only 11 people, as the mines were closed.

Eagle Valley was first settled in the early sixties and Pine Valley somewhat later.[3] The placer deposits of Sparta early attracted attention, but little could be done with them on account of the lack of water until the completion of the Sparta ditch in 1873.[4] During the height of the placer mining Sparta is reported to have been a town of some size and importance. About 1885 valuable mineral discoveries were made near Cornucopia, and since then lode mining has taken precedence over placer mining. The total production of the mines of the Wallowa region through 1923 is thought to be nearly $10,000,000, mainly in gold.

STRATIGRAPHY AND PETROLOGY

Summary Note

The known facts regarding the different kinds of rocks in the Wallowa region and their mutual relations are summarized below. Several subdivisions of the rocks that do not correspond to those recognized in neighboring regions have been mapped and described. As further subdivision will be required when more detailed study is undertaken and many questions remain as to the geologic age of many of the units, few formation names are here employed.

The stratified rocks of supposed Paleozoic age are described as black slate, Clover Creek greenstone, and Carboniferous (?) sedimentary rocks. The black slate is a lithologically distinctive rock several small exposures of which were mapped. Its stratigraphic relations are obscure. The beds assigned to the Clover Creek greenstone are clearly to be correlated with a formation of that name which is widespread in the surrounding region and is of Permian age. The Carboniferous (?) rocks are probably in part younger, in part older than the Clover Creek greenstone. Available data, however, do not permit adequate subdivision.

The Mesozoic stratified rocks, similarly, have been subdivided as far as present information permits. The Martin Bridge formation is a well defined unit, mainly composed of

---

[3] The resources of eastern Oregon, prepared by First Eastern Oregon District Agricultural Society, pp. 45, 49, 1892.
[4] Lindgren, Waldemar, Op. cit., p. 737.

(Photo by Lawrence Panter)
Plate 3-A. Crater Lake. This lake fills a depression made by glacial scour in granodiorite. View looking northeast from Truax Mountain.

(Photo by Jacques Heupgen)
Plate 3-B. Two Granites and Simmons Mountain from the plateau to the east. Note contrast in topography.

Plate 4-A. East side of the valley of upper Clear Creek. Effects of glacial erosion of rock of the diorite-gabbro complex in the plateau.

Plate 4-B. Park at Fish Lake Ranger Station. A typical park in the "Plateau Area".

limestone, of Upper Triassic age. The strata grouped as Triassic (?) volcanic rocks may be stratigraphically equivalent to it, although lithologically very different. The younger Mesozoic sedimentary rocks include strata younger than any of the above and will doubtless be divided into several formations when more information is available.

A number of different intrusive rocks, mainly of Mesozoic age, are recognized and the larger masses have been mapped. The principal body is a stock of granodiorite and related rocks which is extensively exposed in the high mountains. Several other masses, mainly composed of more calcic rocks, are present. Their age relations to the granodiorite are not entirely clear.

The Columbia River basalt forms part of a thick unit of Tertiary age composed mainly of basaltic lava which covers large parts of Oregon and adjacent states. In the Wallowa region, as in many other localities, it contains intercalated clastic beds of different kinds. Dikes, which mark feeders to the flows are exposed in places. Several kinds of unconsolidated materials of Pleistocene and later age are present.

Paleozoic Stratified Rocks

BLACK SLATE

Character.—On East Pine Creek, where it is typically exposed, the black slate consists of black, thin bedded, slate or shale with poorly developed slaty cleavage. The bands are alternately hard and soft. All appear to contain carbonaceous matter, and probably also calcareous material and some are quartzitic. The small area of similar rock farther west comprises brown slate that grades upward into black slate.

The thickness has not been measured but is probably of the order of several hundred feet. On East Pine Creek, the only place where the rocks are well exposed, they are bounded on both sides by Columbia River basalt, and the beds are notably disturbed by minor faulting.

The rock on the east side of Cliff River is black, very fine grained, and has imperfect slaty cleavage. Part of it is somewhat irregularly banded. It was originally a carbonaceous shale.

Age.—If the interpretation of the structure is correct the black slate near East Pine Creek dips under the Carboniferous (?) sedimentary rocks in a syncline whose axis is approximately shown in Fig. 2. This would indicate that the slate was as old or older than any of the other sedimentary rocks in the region.

The black slate near Cliff River clearly dips under and is therefore older than the Clover Creek greenstone, believed to be of Permian age. Both greenstone and slate dip about 45° E., but the strike of the greenstone is nearly north while the slate strikes about N. 10° W. There may consequently be a small angular unconformity between the two formations.

The similarities in lithology and general stratigraphic relations make it probable that the three masses of slate here described have about the same age, although positive correlation is impossible. They are the oldest stratified rocks in the Wallowa region with the possible exception of part of the Carboniferous (?) sedimentary rocks in the eastern plateau area, whose age cannot be determined with the available data. The black slate of the three localities mentioned is thus to be regarded as Carboniferous or older.

CLOVER CREEK GREENSTONE (PERMIAN)

Definition and Distribution.—One of the principal formations in the Wallowa region is a thick sequence of metamorphosed rocks, mostly ancient lava flows and clastic rocks of volcanic origin, but including a little limestone, quartzite and conglomerate. This unit is clearly the same as that named the Clover Creek greenstone by Gilluly.[5] He has mapped the formation in the Baker quadrangle to within a few miles of exposures in the Wallowa region and in a previous paper[6] showed that the same rocks extend east into the region described in the present report. There seems, therefore, no doubt as to the correlation and the name proposed by Gilluly is here adopted.

The principal occurrence of the Clover Creek greenstone in the Wallowa region is in and near the valley of the Imnaha River. The sides of this valley from above the mouth of Cliff

---

[5] Gilluly, James, Geology and mineral resources of the Baker quadrangle, Oregon. U. S. Geol. Survey Bull. 879, p. 21, Pl. 1, 1937.
[6] Gilluly, James, Copper deposits near Keating, Oregon. U. S. Geol. Survey Bull. 830, Pl. 1, pp. 6-13, 1933.

River to an undetermined distance below Indian Crossing are composed of the greenstone and it extends southward underlying the surface of much of the highest parts of the plateau area. The boundary is irregular and except for a short stretch on the west is formed, where the greenstone underlies sedimentary beds, by the overlapping edges of the Columbia River basalt and by intrusive diorite-gabbro complex. East of Russel Mountain the relations are obscured by glacial debris and contacts have not been mapped. In the valley of the Imnaha the greenstone probably continues eastward until it disappears under the Columbia River basalt near the middle of T. 6 S., R. 47 E. A small area of amygdaloidal greenstone similar to some of that near the Imnaha River is exposed on the plateau west of the upper part of Little Eagle Creek. Another irregular area of metamorphic volcanic rock is found in the southwestern part of the region principally in the valleys of Eagle and Little Eagle creeks. This differs in some respects from the greenstone of the valley of the Imnaha but is thought to belong to the same formation.

Character.—Most of the formation is altered lavas that are largely andesitic and dacitic. Associated with them are subordinate amounts of clastic rocks, some of which are of volcanic origin. A few small lenses of limestone were also found.

Near the Imnaha River most of the lavas are different shades of green; others are gray, bluish gray, purple and black. Some are amygdaloidal, many are porphyritic, having in places numerous feldspar phenocrysts as much as an inch or even more in length in dense groundmass. Probably these large, forked phenocrysts result in part from accretion after the consolidation of the rock rather than from original crystallization. Some of the rocks are exceedingly fine grained and free from phenocrysts. In some places there are flow breccias. Conglomerate was noted in several localities, especially on the plateau south of the river, and there are also thin bedded rocks which are probably more or less tuffaceous. Near the head of east Pine Creek a mass of limestone about 250 feet long with a maximum width of some 75 feet is interbedded in the greenstone. It is a fine granular rock that generally resembles the limestone on Holcomb Creek described below. Similar small masses of limestone probably occur elsewhere in the greenstone of this vicinity. Fragments were noted near the head of one of the branches of Clear Creek, but none was found in place.

Owing to the considerable metamorphism in many places it is now impossible to determine strike and dip of these rocks. In such a volcanic complex there are, of course, many dikes belonging to the same period of igneous activity. Some of these can still be easily distinguished, but in many outcrops it is impossible to be sure whether the rock is a flow or a fine grained intrusive. Some of the porphyries crowded with phenocrysts may be irregular intrusions.

The old lavas that form the principal part of the formation in the vicinity of the Imnaha River were originally andesites and dacites. They have all been altered, partially recrystallized and in a few places are schistose. Ferromagnesian minerals do not appear to have been abundant and they have now been altered, mainly to epidote and chlorite. The feldspars not completely clouded by alteration products appear to be oligoclase and oligoclase-andesine. Only a few of the rocks appear to have originally contained quartz. A few of the rocks are nearly black. Some of these have numerous small vesicles filled with calcite. Where the calcite has been leached out there is a superficial resemblance to the far younger Columbia River basalt. These dark rocks have been so thoroughly recrystallized that the original minerals cannot be determined, but rocks are thought to have had the composition of basic andesites.

The metamorphosed volcanic rock from the southwestern part of the region is mainly a fine, even grained dark green, altered andesite which has been somewhat sheared and chloritized. It consists essentially of oligoclase-andesine, and chlorite, with some calcite, epidote, and serpentine. There appear to be no original ferromagnesian minerals remaining. Amygdaloidal greenstone[7] is reported from this area, but the porphyritic rocks characteristic of the greenstones of the Imnaha are rare or absent.

In the greenstone area on Eagle Creek above Holcomb Creek there are several obscure outcrops of rusty quartzite and conglomerate. On Holcomb and Little Eagle creeks lenses of lime-

---
[7] Swartley, A. M., Op. cit., p. 120.

Plate 5-A. The upland surface in "Powder River Valley" near Pleasant Ridge School.

Plate 5-B. View west in Pine Valley from near Sunset. Note terrace in foreground and sharp contrast in the appearance of the valley and its enclosing hills.

stone appear to be interbedded with the greenstone. There is a little similar limestone in the talus on the north side of Eagle Creek below its confluence with Little Eagle Creek, but none was found in place here. The outcrop on the east side of Holcomb Creek consists of hard, blue limestone, brecciated and somewhat recrystallized. On Little Eagle Creek the unaltered limestone is very similar in appearance to that of Holcomb Creek. A considerable portion, however, has been rendered porous by solution and stained with limonite. Bedding can easily be distinguished in the less altered portions of the mass, but there has been thorough brecciation. As a consequence the beds strike and dip irregularly.

Most of the greenstone near the head of Sullivan Creek is green, but locally it is purplish. It is an altered andesitic rock with a dense groundmass consisting of andesine, chlorite and epidote in an almost opaque matrix, dotted with small andesine phenocrysts.

Gilluly[8] notes that the greenstones southwest of the Wallowa region include a large proportion of keratophyric rocks in which the albite is largely secondary. Such rocks were not recognized during the investigation on which the present report is based but it seems probable that they may extend into the region here described.

Thickness.—Here as in neighboring area, the Clover Creek greenstone is so poorly exposed and so obscurely stratified that the thickness of the formation cannot be accurately measured. It is certainly several thousand feet thick. Laney[9] estimates the similar and probably contemporaneous greenstone in Snake River Canyon near Homestead to be 3,000 to 4,000 feet thick. Gilluly[10] estimates that the Clover Creek greenstone in the Baker quadrangle is at least 4,000 feet thick, possibly much greater.

---

[8] Gilluly, James, Copper deposits near Keating, Oregon: U. S. Geol. Survey Bull. 830, pp. 6-8, 1933.

..............., ..., Keratophyres of eastern Oregon and the spilite problem. Amer. Jour. Science 5th series, Vol. 29, No. 171, pp. 225-252, 1935.

..............., ..., Geology and mineral resources of the Baker quadrangle, Oregon: U. S. Geol. Survey Bull. 879, pp. 21-26, 1937.

[9] Laney, F. B., Personal letter of Oct. 17, 1921.

[10] Gilluly, James, Geology and mineral resources of the Baker quadrangle, Oregon: U. S. Geol. Survey Bull. 879, pp. 21-22, 1937.

Age.—The only direct evidence as to the age of the Clover Creek greenstone obtained during the present investigation is afforded by the fact that it underlies beds belonging to the Carboniferous (?) sedimentary rocks, as that term is here used. This is well shown in the neighborhood of the Imnaha River, particularly near Blue Creek, and also near the head of Sullivan Creek farther to the southwest. Near Blue Creek, where the relations are particularly clear, the base of the sedimentary formation consists of green sandstone and conglomerate, both of which appear to have been formed by erosion of the underlying greenstone. The conglomerate here contains abundant greenstone pebbles and resembles much of that interbedded with the greenstone. In some places, as noted below, the Carboniferous (?) rocks have small amounts of volcanic rocks interbedded with them. This and the fact that no evidence of angular unconformity has been observed suggest that the difference in age between the two formations is not great.

Gilluly[11] has found Permian fossils in limestone associated with the Clover Creek greenstone in T. 7 S., R. 41 E., a short distance west of the area covered by the present report. These appear to establish the age of the formation beyond question. It may also be noted that limestone intercalated in the similar greenstone in Snake River Canyon in the general vicinity of Homestead, Oregon, has yielded fossils determined by G. H. Girty to be of the age of the Phosphoria formation of Permian age in eastern Idaho and Montana.[12]

CARBONIFEROUS (?) SEDIMENTARY ROCKS

Definition and Distribution.—All the predominantly sedimentary strata stratigraphically below the Martin Bridge formation (Upper Triassic), except the black slate already described and the sedimentary rocks intercalated in the greenstone series, are here included under the term Carboniferous (?) sedimentary rocks. The great mass of rocks so mapped clearly comprises several formations, but they cannot be separated on the basis of the available data.

---

[11] Gilluly, James, Copper deposits near Keating, Oregon: U. S. Geol. Survey Bull. 830, p. 13, 1933.

[12] Laney, F. B., Personal letter of October 19, 1921. Also personal communication from G. H. Girty.

These rocks crop out in irregular areas in several parts of the region. They are found on Eagle and East Eagle creeks, on the plateau east of these, south of Imnaha River on both sides of Cliff River, and near Cornucopia, on lower Clear Creek, and in the eastern part of the plateau area along Clear, Trail, and East Pine creeks.

Character.—The rocks of this group are all somewhat metamorphosed, the degree of alteration varying in the different localities. There is much jointing, but nowhere has pronounced schistosity been developed. The rocks are all more or less recrystallized and filled with chlorite, mica, and other alteration products. In many places this alteration and recrystallization has gone so far as to obscure primary textures greatly and to render determination of the original character of the rock difficult, and even in extreme instances impossible. Most of the rocks are colored various shades of green and purple, but red is common, and other colors are also found.

The sedimentary rocks overlying the greenstone on Blue Creek are nearly all green sandstone and conglomerate not markedly metamorphosed. On the whole the sandstone is rather coarse and much of it is strikingly cross-bedded. Some fine grained and cherty quartzitic beds are also found. The conglomerate is more abundant near the base of the series. It consists of subangular to rounded pebbles principally from the underlying greenstone in a green sandy matrix. The pebbles are one to four or more inches in greatest diameter. There are a few narrow layers of black trap that may be contemporaneous lava flows interbedded with the sandstone. Similar rocks extend west from here in the mountains on the south side of the Imnaha River. These strata are cut by numerous dikes that include aplite and granitic porphyry related to the nearby stock and diabase and basalt of supposed Tertiary age. Still older dikes may also be present.

The rocks in the Eagle Creek drainage area are for the most part sandstone, conglomerate, and cherty slate with a few strata of probable igneous origin. They are cut by numerous dikes and small irregular masses of fine to medium-grained igneous rock which have shared in the metamorphism of the inclosing strata. Most of the pebbles of the conglomerates are of igneous rock, chiefly fine-grained, porphyritic and anygdaloidal greenstone origin. A few pebbles of gneissic granitic rock were noted, and pebbles of several kinds of metamorphosed sedimentary rocks are also present. Some of the pebbles of the conglomerates are so angular and poorly sorted that they might be breccias. Micaceous sandstone and quartzite with the bedding obscured by metamorphism make up a large part of the series. These and the conglomerates are of various colors, but dark greens and purples predominate. There are also light green and red cherty rocks with imperfect slaty cleavage. These are most common in the hills south of the area of greenstone near the head of Sullivan Creek.

The old sedimentary rocks in the Eagle Creek drainage basin have many similarities with those near the Imnaha River. Specimens of the fine-grained, green, cherty rocks from the two localities could scarcely be distinguished from one another. The conglomerates with angular pebbles that are conspicuous along tributaries of Eagle Creek resemble those in the western part of the area south of the Imnaha. Although discontinuity of surface exposures prevent tracing of one into the other, there is little doubt that the rocks of the two localities are closely related.

On the dissected plateau in the eastern part of the Wallowa region another large body of broadly similar old sedimentary rocks is exposed. Most of the beds are composed of coarse, green, locally cross-bedded, sandstone that somewhat resembles the cross-bedded sandstone on Blue Creek. Conglomerate is present in several places and some of it resembles the conglomerate with angular pebbles in the old rocks in the Eagle Creek drainage basin. The pebbles in some of these beds include igneous rocks like those of the Clover Creek greenstone.

In the middle of the south half of T. 6 S., R. 46 E., there are beds that differ in some respects from any noted elsewhere in the region. These include red and white chert or quartzite, red and green shale and brownish red and green grit with some conglomerate, listed in probable ascending stratigraphic order. Some of the green grit contains large angular fragments of red quartzitic and argillaceous rocks like those which form some of the underlying beds. These beds appear to crop out along an anticlinal axis (Fig. 2) and, therefore, to be stratigraphically lower than those described in the previous paragraph, which flank them on the north and west.

In Sec. 3, T. 6 S., R. 46 E., there is a small outcrop of white marble, whose relations to other rocks are concealed by glacial deposits. An analysis by J. G. Fairchild, U. S. Geological Survey, shows that the marble contains 32.84 per cent lime, 20.52 per cent magnesia, 46.60 per cent carbon dioxide with traces of iron oxides and water, a total of 99.96 per cent. The maximum index of refraction as determined by E. S. Larsen, U. S. Geological Survey, is 1.678. This is slightly less than the corresponding index for pure dolomite which is in accord with the excess of lime shown by the analysis.

Thickness.—No satisfactory place for measurement of the thickness of the Carboniferous (?) sedimentary rocks has been found. Along Blue Creek and near Cliff River, as shown in section AA^1, Plate 1, it is somewhat more than 1,000 feet but can hardly be much more than 2,000 feet. In the Eagle Creek drainage basin, especially near Sullivan Creek, the thickness may be somewhat greater. The structure here is so complex that it is difficult to estimate the thickness of the unit. In the plateau in the eastern part of the region, the thickness must be great. Section AA^1, Plate 1, suggests that it is much in excess of 10,000 feet. It seems possible that structures not yet detected may cause some duplication of beds. Nevertheless, the broad expanse of beds with moderate dips shown in the eastern part of Plate 1 appears to imply that the thickness of the unit in this part of the region is far greater than it is in any of the exposures in the high mountains farther west.

Age.—The old sedimentary strata in the high mountains appear to overlie the greenstone series without angular unconformity and may merely record a shift in conditions of deposition. The old sedimentary rocks in different places lie below either the Martin Bridge formation or volcanic rocks that may be contemporaneous with it. Unconformity with the Martin Bridge formation and relatively greater deformation in the older rocks are well shown on Paddy Creek and near the head of Spring Creek and less definitely elsewhere. Where the Triassic (?) volcanic rocks overlie the old sedimentary strata the relations are less clear but on the east side of Red Mountain the contact is irregularly rolling and appears to be an erosional unconformity.

In and near Ts. 6 and 7 S., R. 46 E., where the Carboniferous (?) beds are thicker and in part lithologically different from those in the high mountains, the stratigraphic relations are less definitely known. The presence of greenstone pebbles in some of the conglomerate on the plateau suggests that these beds, at least, are younger than the Clover Creek greenstone like the similar beds in the mountains. On the other hand, the sedimentary rocks southeast of the diorite-gabbro complex are so placed with reference to the greenstone north of that intrusion as to suggest that they are stratigraphically equivalent to or may even underlie the greenstone. The structural conditions here are discussed on pages 74 and 75. It may be significant in this connection that the cherts and black slate that are the sedimentary rocks on the plateau that are most strikingly different from those in the mountains are both apparently stratigraphically low. They may belong to a part of the Carboniferous (?) sedimentary rocks not yet uncovered by erosion in the high mountains. The possibility that part of the Carboniferous (?) sedimentary rocks may be similar in age to the Clover Creek greenstone is supported by the fact that where the relations are comparatively well exposed no unconformity has been detected and by the presence within the mapped greenstone of conglomerate and other beds lithologically indistinguishable from some of the beds included in the Carboniferous (?) sedimentary rocks. Also the sedimentary rocks in a few places include layers of igneous rock thought to represent old lava flows such as those that make up most of the greenstone.

The close relations between the Clover Creek greenstone and the Carboniferous (?) sedimentary rocks strongly suggests that much of the latter is of Permian age. Wherever they are in contact with the Martin Bridge formation it is evident that the Carboniferous (?) rocks are much the older. As nothing has been observed which shows that any part of this unit is greatly older than the greenstone the whole of it may be tentatively regarded as of Carboniferous age. This is in accord with such data as are available in regard to Paleozoic rocks in neighboring areas. Paleozoic fossils have been found in several localities in eastern Oregon but without exception those so far reported[13] have been assigned, with diverse degrees of certainty, to the Carboniferous.

EPIDOTE-GARNET ROCK

Definition and Distribution.—On the ridge between Cliff River and Blue Creek there is a peculiar green rock so metamorphosed as to obscure its original character. It is now predominantly an epidote-garnet rock.

Character.—This epidote-garnet rock is a lithologic, not a stratigraphic, unit, and is derived from the metamorphism of rocks of possibly diverse origins. Garnet appears more abundant on the Cliff River than on the Blue Creek side of the mass. Locally on the Cliff River side much of the rock is an altered conglomerate whose pebbles are sub-angular to rounded fragments of fine grained igneous rock. Except that there is much epidote, some in well formed crystals, the rock resembles the conglomerate farther down the slope that belongs to the group of older sediments just described. Indeed no sharp line of demarcation can be drawn. There is complete gradation between rock dominantly composed of epidote and conglomerate in which epidote is only sparingly present. Some talus fragments of the epidotized rock consist entirely of metamorphosed basic igneous rock, originally a basalt or pyroxene andesite. On the Cliff River side most of the rock appears to be greenstone of different kinds.

---

[13] Lindgren, Waldemar, The gold belt of the Blue Mountains of Oregon, 22nd Ann. Rept. U. S. Geol. Survey Part 2, p. 578, 1902.

Washburne, Chester, Notes on the marine sediments of eastern Oregon, Jour. Geol. Vol. 11, pp. 224-225, 1903.

Packard, E. L., A new section of Paleozoic and Mesozoic rocks in central Oregon. Amer. Jour. Sci. 5th ser., Vol. 15, pp. 221-224, 1928.

Gilluly, James, Copper deposits near Keating, Oregon: U. S. Geol. Survey Bull. 830, pp. 6-13, 1932.

Gilluly, James; Reed, J. C.; and Park, F. C., Jr., Some ore deposits of eastern Oregon: U. S. Geol. Survey Bull. 846.

Mesozoic Stratified Rocks

MARTIN BRIDGE FORMATION

Definition and Distribution. — The calcareous formation which is so prominent in the western part of the region is here named the Martin Bridge formation, from the bridge across Eagle Creek, near which the best preserved fossils were found. The formation borders Eagle Creek from a point about a mile below Martin Bridge to an undetermined point beyond the confluence of East Eagle Creek, and extends up the valley of East Eagle Creek and across the divide to the head of the Imnaha River. Prominent outcrops of the limestone can be seen far to the north of the region mapped. West of Eagle Creek outcrops were noted near the Sparta Ditch and on one of the forks of Goose Creek. Other outcrops are found to the east on Little Eagle and Spring creeks. The metamorphosed lava in the valley of Cliff River and on Red Mountain may be the stratigraphic equivalent of the Martin Bridge formation in this locality as is more fully discussed below.

Character.—Most of this formation is limestone, which in most places has been converted into a marble, which is especially conspicuous near granitic masses near Martin Bridge on Eagle Creek and along the lower course of Paddy Creek the formation is less metamorphosed than it is farther north. Here the lower part consists of limy shale and impure limestone, with a few greenish tuffaceous beds, and near the base, a little sandstone and conglomerate. The thickness of this argillaceous member is variable, in places probably about 100 feet. Above it is a massive blue limestone. The lower part of this is a breccia which on weathered surfaces, looks like conglomerate. On polished surfaces, however, the structure is clearly seen to result from fracturing of the blue limestone. The resemblance to a conglomerate is increased by the fact that shaly beds are unfractured, so that the seeming conglomerate locally appears to be interbedded with shale. The upper part of the formation consists of black, thin-bedded limestone with beds of limy shale. Some of the limestone in the formation contains lenses of black chert.

The thicknesses of the different members of the formation varies, but they have nowhere been accurately measured. The total thickness of the formation at this place certainly exceeds 1,000 feet. Fossils are locally abundant in the shaly beds but they are not well preserved.

The formation is almost continuously exposed from Martin Bridge to East Eagle Creek where marble crops out extensively. Although any fossils which may have existed in the marble have been obliterated by the metamorphism, and the subdivisions of the formation described above cannot be recognized with certainty, there seems to be no doubt of the correlation. In general the marble is a massive blue-gray crystalline rock in which stratification is indistinct or lacking. On Hudson Creek, and elsewhere, it contains beds of impure limestone and limy shale. Near the mouth of the canyon of Hudson Creek some angular blocks of fine-grained green metamorphic rock are included in the marble. At this locality the marble has close partings and may correspond to the thin-bedded lower member of the formation near Martin Bridge. The upper part of the section as exposed along this creek contains calcareous schist interbedded with lenses of crystalline limestone. This may correspond to the upper member near Martin Bridge. Just north of Hudson Creek, a conspicuous mountain is composed of blue-gray marble with several white bands so distorted and crumpled as to show that pressure has been severe. The thickness of the formation as exposed in Hudson Creek seems to be much more than 2,500 feet, but it is so much crumpled that this may be excessive. On the east side of East Eagle Creek opposite Hudson Creek and the marble mountain, it is irregular but apparently considerably less than 2,000 feet thick at its maximum.

Near the head of Imnaha River this formation forms the mountains on both sides of the stream. It can be definitely traced into the marble on East Eagle Creek and there is no doubt of the correlation. Most of the rock here is a compact blue-gray marble. Bedding is visible in many places, but much of the rock is massive and without visible stratification. On the downstream side at the base of the marble mass there are beds of impure black micaceous limestone and limy shale, with minor amounts of sandstone and conglomerate. Some of the limestone contains numerous elongate, iron-stained concretions. These beds probably correspond to the lower member near

Martin Bridge. Even though fossils are present they are very poorly preserved. Similar beds of dark-impure limestone with associated quartzite occur in the marble farther upstream, but their place in the section was not determined. The total thickness of the formation here appears to be more than 2,500 feet, and may be as much as 3,000 feet if there is no thickening by crumpling or faulting.

Near the head of Imnaha River the marble is intruded by a batholithic mass of granodiorite and related rock which extends far to the west. Sketch corrections of the topography are shown in this locality on Plate I, but the map is still so generalized that the geologic contacts here cannot be accurately shown. The granitic rock has metamorphosed the marble nearby. In the basin at the head of the Imnaha the marble is irregularly recrystallized so that some of the individual carbonate grains are more than two inches across. Close to the contact the igneous rock has absorbed considerable amounts of calcium and in places is converted into amphibolite. In several areas the marble has been replaced by epidote, garnet, and other minerals. In the east half of sec. 24, T. 5 S., R. 44 E., specimens of tremolite, and of silicified marble containing crystals of scapolite and many small crystals of pyrite were found in the talus.

Age.—The Martin Bridge formation is the only pre-Tertiary formation in the region mapped that has yielded indentifiable fossils. Lindgren[14] long ago collected fossils from the limestone on Eagle Creek one-third mile below the mouth of East Eagle Creek. These were reported by T. W. Stanton to consist of numerous specimens of *Halobia* and two undeterminable fragments of an ammonite. From the Miles placers, one and one-half miles below the mouth of East Eagle Creek, Lindgren obtained through Mr. F. R. Mellis, of Baker, Oregon, a cast of a gigantic gastropod found in the limestone. Doctor Stanton remarked that it had the form of a very large *Turritella* or *Pseudomelania*. Another lot of fossils collected by Lindgren from the limestone two and a half miles above the mouth of East Eagle Creek contained *Pentacrinus* columns with spines and fragments of tests of echinoids.

---

[14] Lindgren, Waldemar, Op. cit., p. 581.

At Martin Bridge a partial section of this formation, totaling 630 feet, has been measured by J. P. Smith[15]. His section is given below.

SECTION ON EAGLE CREEK, BAKER COUNTY, OREGON

Upper Triassic	Thickness (feet)
Massive limestone without visible fossils	60
Dark brown argillaceous shales, with *Halobia* cf. *austriaca*, and other species of *Halobia*, and *Daonella*?	100
This thin-bedded limestone, with banks of corals, *Thecosmilia norica* Frech, *Spongiomorpha* cf. *acyclica* Frech, *Montlivaultia norica* Frech, *Heterastridium conglobatum* Reuss	40
Barren shales	300
Massive limestone without fossils	100
Calcareous shales, with *Halobia*, cf. *superba*, *H.* cf. *salinarum*, *H.* cf. *austriaca*, *Dittmarites* sp.? etc.	30 visible

Fossils collected by the writer from impure limestone at several localities have been examined by Doctor T. W. Stanton. In a lot collected on Eagle Creek below Martin Bridge, *Terabratula?* sp. *Halobia* sp. related to *H. superba* Majsisovics, and other pelecypods represented by imprints were found. Another lot collected in the same vicinity contains *Halobia* sp. related to *H. superba* Majsisovics, *Clionites* sp., and an undetermined ammonoid. These two lots are definitely referred to the Upper Triassic by Doctor Stanton. Two lots were collected from the impure limestone near the base of the Martin Bridge formation on the south side of Imnaha River. The one collected in the SW. ¼ sec. 19, T. 5 S., R. 45 E. "contains imperfect prints of two forms of Pectinoid shells". That collected in the NW. ¼ sec. 20, T. 5 S., R. 45 E. "contains a number of very imperfectly preserved coiled forms which seem to be ammonites possibly belonging to the genus *Arcestes*". Doctor Stanton considers that the latter can be assigned without much question to about the same position as the two lots from near Martin Bridge. The field relations show clearly that the two collections from the valley of Imnaha River come from essentially the same horizon. A collection made in the SE.¼ sec. 6, T. 7 S., R. 45 E. "contains

---

[15] Smith, J. P., The occurrence of coral reefs in the Triassic of North America, Am. Jour. of Science, 4th Series, Vol. XXXIII, No. 194, pp. 94-95, 1912.

fragmentary gastropods of a naticoid type, possibly very imperfect young ammonites, and fragments of bone". Another collection from the same mass of limestone in the NE.¼ sec. 7, T. 7 S., R. 44 E. "contains some undetermined corals". These two collections "contain nothing recognized as distinctly characteristic and are doubtfully referred to the Triassic". The lithology of the rocks supports the view that they belong to the same formation as the limestone of East Eagle Creek. The Martin Bridge formation can thus be confidently referred to the Upper Triassic.

TRIASSIC (?) VOLCANIC ROCKS

Definition and Character.—Certain beds in which volcanic material predominates are exposed along Cliff River. From their field relations, outlined below, these rocks here named the Triassic (?) volcanic rocks, are believed to be the stratigraphic equivalent of the Martin Bridge formation. They occupy part of the valley of Cliff River and their thinner eastward extension crops out on the east side of Red Mountain west of the head of Norway Creek.

Character.—These rocks have undergone both dynamic and contact metamorphism, and are so intensely altered that their original character is almost unrecognizable in the field. They are mostly black or dark bluish gray, fine-grained, and distinctly stratified. In some exposures there are well-defined layers a few inches thick. Between the bands of dark rock there are occasional small lenses of rather coarsely crystalline calcite, in part silicified. These are rarely more than a few inches long. Numerous small quartz veins cut the rock especially near granitic contacts and there are nodular masses of green epidote and quartz. Locally the rock appears to be largely replaced by epidote. There are small dikes of metamorphosed trap cutting the layered rock. Some of these are approximately parallel to the layers with tongues only fractions of an inch in diameter extending out from them across the layers. Specimens of the dark rock from Cliff River and from the similar exposures on the east side of Red Mountain have been examined petrographically and proved to be fine-grained igneous rocks. They contain notable amounts of calcite, epidote, garnet, chlorite, and other products of metamorphism. The feldspars are altered and cannot be precisely identified, but are principally plagioclase. Biotite is present in some, and augite in others. The rocks were probably originally andesites and basalts.

These igneous rocks do not cut the sedimentary rocks with which they are in contact nor do they show any other evidence of intrusive relations. On Red Mountain the contact with the younger sedimentary rocks above is nearly or quite conformable, and the base of the stratified igneous rocks rests in apparent erosional unconformity on Carboniferous (?) beds. Thus it is probable that the igneous rocks are extrusive lavas. Some of the more thinly bedded rocks may be tuffaceous. The calcareous masses associated with the volcanic rocks presumably represent lenses of limestone originally laid down with them.

Age.—In so far as the stratigraphic position of the volcanic rocks has been established they are equivalent to the Martin Bridge formation exposed farther west. The sedimentary strata below them have been shown to correspond to those underlying the Martin Bridge formation, and the relations in both cases appear to be unconformable. As will be shown below, the rocks overlying the volcanic group are believed to be the equivalent of the formation overlying the Martin Bridge formation. If the interpretation of the structure adopted in this report is correct, the stratigraphic equivalent of the Martin Bridge formation must outcrop in the situation occupied by these volcanic rocks. The volcanic rocks have intercalated calcite lenses that suggest conditions of sedimentation during the period of volcanism similar to those under which the Martin Bridge formation was laid down. On the other hand, the tuffaceous beds known at several localities in the Martin Bridge formation, prove the presence of volcanism during the deposition of that formation. From the above considerations it is believed that the volcanic group is essentially contemporaneous with the Martin Bridge formation. It seems probable that the only real distinction between them is that at the locality of the volcanic rocks, volcanism was more active, and such thick calcareous sedimentary deposits did not form as was the case farther west in the locality of the more typical phases of the Martin Bridge formation. The age of the volcanic rocks is, therefore, considered to be probably Upper Triassic.

YOUNGER MESOZOIC SEDIMENTARY ROCKS

Definition and Distribution.—Overlying the Martin Bridge formation are thick sediments of several varieties. These have not yet been differentiated in mapping and the whole sequence is grouped as younger Mesozoic sedimentary rocks. They are mapped on the east side of Humming Bird Mountain, on Krag Peak and the ridge to the north, on Red and Granite mountains, and in Twin Canyon. Certain patches of shale above the limestone on Eagle Creek, so small that they were not differentiated in mapping, may be of the same age. Likewise the inclusions of hornfels in the granitic stock near Cornucopia may belong to the younger Mesozoic rocks. However, this hornfels near Cornucopia is in unconnected slivers and irregular fragments in the igneous rock, and positive correlation between them and other sedimentary masses cannot be made.

Character.—The lithology of these rocks varies in different localities. On Eagle Creek near Martin Bridge, and in the hills just northeast, soft brown more or less carbonaceous shale overlie the Martin Bridge formation. The shale outcrops are very small and were not mapped separately. No stratigraphic break was detected between it and the limestone, but the shale is believed to be the unmetamorphosed equivalent of some of the younger Mesozoic strata farther north. Similar shale rendered schistose by pressure is interfolded with the limestone along the upper part of Gold King Creek.

Near the head of Hudson Creek on the east side of Humming Bird Mountain thick metamorphosed sedimentary rocks are exposed stratigraphically above the overturned beds of the Martin Bridge formation. These rocks are mostly fine-grained sandstone and shale, in part quartzitic, predominantly dark gray in color, somewhat schistose and partly recrystallized. A little conglomerate was noted near the base of the series. This contains pebbles of various metamorphic rocks, of a granitic rock, and subangular blocks of limestone. The rocks have undergone both dynamic and hydro-thermal metamorphism not so intense, however, as to entirely obliterate the stratification. Amphibole, mica and other metamorphic minerals have been formed in variable but usually considerable amounts. Calcite veins fill many of the joint planes. Many dikes of diorite and related rock, apophyses of the granitic rock of Humming Bird Mountain, cut the sediments in various directions. Emanations from these probably contributed to the metamorphism of the rock. Some of the talus here contains metamorphosed fine-grained igneous rock that probably indicates the presence of earlier dikes that shared in the metamorphism of the sedimentary strata. Unmetamorphosed diabase dikes, probably of Tertiary age, also occur. Some of these are large and form prominent features of the scenery. A little mocaceous metamorphosed sandstone, like some of that just described occurs in Twin Canyon.

On Krag Peak and the ridge to the north is a thick sequence of slate, micaceous and quartzitic sandstones and conglomerate overlying the Martin Bridge formation to the west. Most of these rocks are black or dark bluish or greenish gray on fresh surfaces. Some are thin, others coarsely and irregularly bedded. These rocks are sheared and recrystallized. They contain metamorphic amphibole and mica like that of the rocks just described. Near the intrusion close to Corral Creek the rock has been epidotized in patches. In general, many of the rocks near Krag Peak resemble in appearance, degree of metamorphism, and probable origin those east of Humming Bird Mountain described above. They also contain the same three groups of dikes. However, conglomerate is more wide spread near the head of Cliff River, east of Krag Peak, than on Hudson Creek and it is lithologically different. It has elongate, rounded to subangular pebbles a fraction of an inch to a couple of inches long. Many of the pebbles are quartose but a few are impure limestone containing round crinoid stems. No epidote like that seen in some of the rocks of Corral Creek was noted on Hudson Creek.

Red Mountain and the ridges to the north and south are evidently made up of the eastern continuation of the rocks just described. These are predominantly black or gray thin-bedded slate and quartzite, strikingly like the finer phases of the rocks of Krag Peak and near Humming Bird Mountain, but the coarse phases appear to be absent and no conglomerate was noted. Dikes of aplite, diorite, pegmatite, and similar rocks are abundant. Diabase and basalt dikes are also present, but the older metamorphosed dikes were not noted.

The thickness of these strata cannot be determined, as the top has not been recognized. On Hudson Creek, probably the best locality for measurements, there is an apparent thickness of about 3,000 feet, but the upper beds are intruded and cut off by granitic rock. The thickness near Krag Peak and Red Mountain is of about the same order.

Age.—The age of these strata is unknown except that they are younger than the Martin Bridge formation on which they lie and older than the granitic rock that intrudes them. The limestone conglomerate on Hudson Creek indicates some erosional break between these strata and the Martin Bridge formation, but this is the only evidence of unconformity known. Any unconformity that may exist is probably not large. No identifiable fossils have been found. Thus the evidence is sufficient to prove that the strata are of Mesozoic age, and later than the Upper Triassic. They are probably largely Jurassic, but may extend into the Cretaceous. That Jurassic sedimentary rocks are present in the general region is proved by the fact that Jurassic fossils have been found in several localities.[16]

HORNFELS

Definition and Distribution.—The dark fine-grained rock on the southeast slopes of Two Granites Mountain near Cornucopia is largely hornfels. It occupies an irregular area of about two square miles, with much more irregular boundaries than could be shown on Plate 1. It is on the contact of the granitic stock (pl. 6-B). The main body mapped is intricately intruded by innumerable irregular masses of granitic rock and dikes of kindred rock together with numerous Miocene (?) basaltic dikes. The different intrusive masses are so abundant that in parts of the area mapped as hornfels there is probably more intrusive rock than hornfels at the surface. Mine workings show slivers and irregular masses of hornfels enclosed in the granitic rock. These have not been distinguished on Plate 1. This rock is present in all the large mines in the vicinity of Cornucopia.

---

[16] Hyatt, A., Trias and Jura in the western states. Bull. Geol. Soc. Amer. Vol. V, p. 401, 1894.

Packard, E. L., A new section of Paleozoic and Mesozoic rocks in central Oregon: Amer. Jour. of Sciences, 5th series, Vol. 15, pp. 221-224, 1928.

Lupher, R. L., Geological section of the Ochoco Range and Silver Plateau south of Canyon City, Oregon. (abstract) Geol. Soc. America Bull. Vol. 42, No. 1, pp. 314-315, March 31, 1931.

Goodspeed[17] has recently suggested that some of the blocks of granitic rock in the hornfels result from recrystallization of the latter.

Character.—The hornfels is a fine-grained rock, nearly black in most exposures but locally gray, and some surfaces are greenish. Layering probably corresponding to the original bedding can be discerned in several exposures, but a large part of the rock exhibits only cleavage which may have no relation to bedding.

The hornfels consists essentially of quartz, oligoclase-andesine, and somewhat chloritized biotite. Green hornblende, zoisite, and black iron ore are present locally in minor amount. The texture and composition show that the rock is clastic but it contains much material derived from igneous rock. It was originally a shale, but has been metamorphosed both by orogenic pressure and by the granitic rock which intrudes and largely incloses it, until it is now hornfels. In addition the hornfels bordering the veins was altered during the mineralization.

The rock here called hornfels is locally termed greenstone. Swartley[18] states that in parts of the Union-Companion mine the wall-rock of the vein is "part of an old intrusion or flow now altered to greenstone". Such an occurrence was not observed during the present examination but many parts of this mine are now inaccessible. In other parts of the mine he found that the rock intruded by the granodiorite was "greenish schist, originally probably a basic sandstone" and in the Last Chance Mine[19] "a dense dark green rock that was probably once an argillaceous sediment laid down between the old surface flows". Both of these are doubtless to be included in the hornfels, as the term is used in this report.

Age.—Direct evidence as to the age of the hornfels is lacking. It is isolated from all other stratified rocks. It is tentatively correlated with the younger Mesozoic sedimentary rocks because it is in strike with these rocks on Red Mountain and resembles them in general, but the evidence is inconclusive.

---

[17] Goodspeed, G. E., Small granodiorite blocks formed by additive metamorphism: Jour. Geol., Vol. 45, No. 7, pp. 741-762, 1937.
[18] Swartley, A. M., Op. cit., p. 49.
[19] Swartley, A. M., Op. cit., p. 49.

The abundance of igneous material in the hornfels suggests that it may belong with the Triassic (?) volcanic rocks. Many of the dark colored rocks of this group are similar in general appearance to the more massive varieties of the hornfels, and some of them may be of clastic origin. If Swartley is correct in the belief that igneous rocks are included in the hornfels the possibility that it belongs to the Triassic (?) volcanic rocks becomes stronger. Still another possibility is that the hornfels is the equivalent of the black slate that underlies the greenstone series on the east side of Cliff River.

Mesozoic and Older Intrusive Rocks

METAMORPHOSED DIKE ROCKS

Definition and Distribution.—Dikes and irregular dikelike intrusions are plentiful in parts of the Clover Creek greenstone and the Carboniferous (?) sedimentary rocks, and also occur in the Martin Bridge and younger sedimentary formations. Many of these are so metamorphosed that it is difficult without close examination to distinguish them from the sedimentary and volcanic strata which they cut. Such rocks are abundant along East Eagle Creek and are present in most parts of the area. No attempt has been made to show them on the map. Nearly all form such small masses that it would be impossible to do so.

Character.—Most of these rocks are in small dikes and irregular tongues from a few inches to a few feet thick. In a few localities as on the north side of Little Kettle Creek near its mouth, the outcrop is several hundred feet wide.

Most of the dike rocks are metamorphosed traps. They are fine-grained, dark colored, and commonly micaceous. They are so metamorphosed as to make exact determination of their original characters impossible.

Age.—Rocks of more than one period of intrusion are represented. Many of the dikes cutting the greenstone are genetically related to the lava in that formation, and hence probably of Permian age. As the rocks cutting the sedimentary strata have undergone metamorphism comparable to that of the enclosing rocks they are probably of approximately comparable age.

AMPHIBOLITE

Definition and Distribution.—The irregular mass of granular igneous rock which extends south along Clear Creek from its junction with Trail Creek may be called amphibolite. This mass and the oval body of similar rock between Clear Creek and East Pine Creek and almost or quite detached from the main body crop out over an area of a little more than three square miles.

Character.—The amphibolite forms an irregular stocklike mass, intrusive into the Carboniferous (?) sedimentary rocks and in part overlapped by the much later Columbia River basalt. It is somewhat more weathered than the granodiorite and quartz diorite described below, and does not stand out in such prominent outcrops.

The rock is medium grained and composed essentially of hornblende with some residual cores of augite, biotite, altered feldspar and subordinate amounts of quartz. Calcite, chlorite, epidote, and other alteration products are common. The feldspar is now albite or albite-oligoclase, but was probably originally more calcic. Before metamorphism the most abundant ferro-magnesian mineral was augite, but this has now largely gone over into hornblende. The rock shows evidence of some crushing or straining but not to a very marked degree. It is an amphibolite resulting from the metamorphism of a quartz diorite or similar rock.

Age.—The field relations of this rock are essentially similar to those of the granodiorite and diorite described below. The original petrographic character was also somewhat similar. The much greater degree of alteration and the fact that there is some evidence of crushing suggests greater age. Probably it was intruded at some time in the early Mesozoic, but the data at hand are not sufficient to date it more closely. The rock forming the small elliptical mass south of the larger body is more weathered and in the hand specimen appears so different from the main body that it was first thought to be older. The appearance under the microscope is, however, so similar that this does not appear probable.

DIORITE-GABBRO COMPLEX

Definition and Distribution.—An intrusive mass crops out along the upper course of Clear Creek and extends east to Fish Lake and Russell Mountain. The composition of the mass varies, and it is best designated a diorite-gabbro complex. It extends a short distance farther east than the area mapped, and the known area is about four square miles.

Character.—This mass is an irregular stock. In few places could the actual contacts with older rocks be observed, but it evidently cuts steeply through the greenstone and sedimentary rocks. It is but little weathered and stands up in prominent ledges, but its relation to other formations is often obscured by a covering of alluvial soil. Only the thickest and most prominent masses of alluvium are shown on the map, plate I. Columbia River basalt laps over its surface on the south side.

The rock varies in appearance as well as composition. Most of it contains so much pyroxene as to be almost black and is finer grained than the average granitoid rock farther west. There are some small patches composed almost exclusively of ferromagnesian minerals. In many localities, however, the rock is coarser and contains a smaller proportion of dark minerals than the average. Locally it much resembles some of the more basic phases of the main stock to the west.

The rock is so variable in texture and composition that it is difficult to select specimens that can be considered at all representative of the whole. Much of the mass consists essentially of titaniferous augite, hypersthene, hornblende, dark green biotite, and bytownite and has marked flow structure. There is a very little interstitial quartz. The rock is fresh with very little development of secondary minerals.

More silicic portions of the intrusive mass consist essentially of quartz, sericitized oligoclase, chlorite, epidote, and subordinate muscovite. The chlorite and epidote are alteration products of pyroxene. No residual pyroxene was found, but its crystal form is preserved in some of the chlorite-epidote aggregates. The rock is thus a somewhat altered quartz-pyroxene diorite.

Many small dikes and irregular intrusions cut this complex and the contiguous stratified rocks. Most of these are genetically related to the complex. Some appear megascopically to be more basic than the gabbro described above and others are less so. Most of them are dark fine-grained porphyritic traps.

Age.—So far as can be judged from data obtained during the present investigation the diorite-gabbro complex is similar in field relations and in absence of conspicuous dynamo-metamorphism to the larger, more silicic masses farther west, described on pages 47-50. On this basis, the different stocks are thought to be broadly of the same age. As they cut rocks at least as young as Permian the intrusions can hardly be older than Mesozoic. Detailed work over a wide area in eastern Oregon has led Gilluly and his associates[20] to the conclusion that rocks lithologically somewhat similar to the diorite-gabbro complex in the eastern part of the Wallowa region are older than those similar to the quartz diorite and granodiorite farther west in that region.

ALBITE GRANITE

Definition and Distribution.—The mass of granitoid rock outcropping in the vicinity of Sparta may in general be termed an albite granite. It is overlapped in most places by Columbia River basalt so that its shape and extent are obscured. North of Powder River it crops out over an area of nearly 40 square miles. On the south side of that stream it is concealed by the basalt flows that cover the hills in that direction.

Character.—The granite is in most places deeply weathered and disintegrated and does not form prominent outcrops. There is considerable local variation in the rock, although the main mass is fairly uniform in composition. It is medium grained with allotriomorphic texture. The essential minerals are quartz and somewhat sericitized albite, the former being more abundant. There is a little biotite almost completely altered to

---

[20] Reed, J. C. and Gilluly, James, Heavy mineral assemblages of some of the plutonic rocks of eastern Oregon: American Mineralogist, Vol. 17, No. 6, pp. 202-204, 1932.

Gilluly, James; Reed, J. C., and Park, C. F., Jr., Some mining districts of eastern Oregon: U. S. Geol. Survey Bull. 846, pp. 17-18, 1934.

Gilluly, James, Geology and mineral resources of the Baker quadrangle, Oregon: U. S. Geol. Survey Bull. 879, pp. 27-29, 1937.

chlorite. A partial analysis of granite from this mass by W. F. Hillebrand[21] is given below.

PARTIAL ANALYSIS OF GRANITE FROM SPARTA

Constituent	Per Cent
SiO_2	76.25
CaO	1.70
Na_2O	4.60
K_2O	.59

In the south, especially along Powder River, there are areas of basic rocks in the granite. Some of these are dikes, but some appear to be differentiation products that grade into the more normal granite. The most extreme example of such differentiation noted is gabbro found in the canyon of Powder River near Black Bridge. It is a medium-grained rock consisting essentially of augite and labradorite, fresher than the granite just described, but with its pyroxene being replaced by chlorite and with the twinning laminae of the plagioclase somewhat distorted as though by pressure. There is considerable pyrite and black iron oxide. Along Powder River several irregular masses of quartzite and micaceous schist rocks are included in the granite. The individual masses are so small and irregular that they are not shown on the map. A few are as much as several hundred feet in thickness.

In the north, near Martin Bridge, there is a fine-grained somewhat variable porphyritic rock that is considered to be a marginal phase of the albite granite. It is a fine-grained greenish much altered felsite with small phenocrysts of quartz and feldspar. The rocks do not seem to differ greatly in composition from the granite into which they appear to grade. A somewhat similar rock forms dikes and irregular intrusions along Eagle Creek both above and below the mouth of Little Eagle Creek. This is a fine-grained porphyritic rock consisting for the most part of recrystallized quartz and oligoclase. It is thus an aplite alaskite or related rock.

The concept has recently been advanced[22] that several masses of albite granite in this vicinity, including the one here de-

---

[21] Lindgren, Waldemar, The gold belt of the Blue Mountains of Oregon: U. S. Geol. Survey 22d Ann. Rept., Part II, p. 586, 1901

[22] Gilluly, James, Replacement origin of the albite granite near Sparta, Oregon: U. S. Geological Survey Prof. Paper 175-C, pp. 65-81, 1933.

scribed, originated by hydrothermal replacement of quartz diorite.

Age.—The albite granite is intrusive into the Clover Creek greenstone and its finer grained marginal phase cuts the Martin Bridge formation. So far as can be judged from the data at hand it seems probable that the albite granite was intruded late in the Mesozoic era and probably at about the same time as the stock near Cornucopia.

In the paper just cited, Gilluly says that the quartz diorite, that he regards as the rock from which the albite granite was derived, is older than such quartz diorite and granodiorite as that near Cornucopia. He suggests that the albite alteration may have taken place during the magmatic period in which the Clover Creek greenstone was erupted. His suggestions are based on study of a number of masses of albite granite in the general region. The relations to Upper Triassic beds shown on plate 1 seem to preclude any genetic relation between the granite near Sparta and Permian igneous activity.

QUARTZ DIORITE AND GRANODIORITE

Definition and Distribution.—Granitic rock of intermediate character is widespread in the Wallowa Mountains. Its composition is variable but most of the rock is either quartz diorite or granodiorite. The mass that contains all the principal mines of the Cornucopia district crops out in an irregular ellipse with the long axis extending north of of west from Cornucopia. The area covered is a little over 8½ square miles. West and north of it are several smaller, irregular shaped masses of similar rock. The total area occupied by these is somewhat over 2 square miles. Much the largest intrusive mass in the region is that which forms Humming Bird Mountain and extends far to the north and west. This mass was studied in the field only on Humming Bird Mountain and on the slopes about the head of Imnaha River. It is clear that the rock in the two localities belongs to the same great massif. From vantage points on neighboring peaks the granitic rock can be seen extending for mile after mile in rough, bare mountain summits. Swartley[23]

---
[23] Swartley, A. M., Ore deposits of northwestern Oregon, The Mineral Resources of Oregon. Vol. 1, No. 8, p. 25, Dec. 1914.

Plate 6-A. The irregular contact of the stock on the West Fork of Pine Creek. The granitic rock is on the left of the picture. The strata on the right belongs to the younger Mesozoic sedimentary rocks.

Plate 6-B. Bonanza Basin between the Last Chance and Queen of the West mines. View, taken from the latter, shows inclusion of hornfels in granodiorite; basalt dike is visible in the distance.

estimates the extent of such rock in the Wallowa Mountains as about 250 square miles.

Character.—The mountains composed of this rock show among the most striking scenic features of the range. They tower above their surroundings and are for the most part nearly bare of vegetation and soil. The rock is in many places mechanically disintegrated to a marked degree, but there has been comparatively little chemical breaking down of the constituent minerals. The smaller areas of the rock generally have somewhat gentler slopes, are better covered with soil and vegetation and are, therefore, much less conspicuous features.

The rock has a medium-grained, hypidiomorphic texture. It is composed of sodic andesine, subordinate orthoclase, quartz, biotite, muscovite, hornblende, and augite with ilmenite, titanite and other minor accessories. The proportions of the different constituents vary within rather wide limits. Locally quartz is abundant and elsewhere it is almost absent. Biotite is uniformly present, usually in notable amount. In most localities black hornblende is also a prominent constituent. Muscovite and augite are found in places but are not widespread. Chlorite and some sericite occur, but in general the minerals have undergone little alteration. The rock ranges in composition from granodiorite to quartz diorite and locally is a diorite. There is much variation in the proportions of the component minerals, especially near the contacts. Some patches along the borders of the intrusive masses consist almost exclusively of hornblende.

Age.—The different intrusive masses in the region composed of rock such as that described above are doubtless genetically related. All that can be definitely stated from the field evidence in the Wallowa region regarding their age is that intrusion took place considerably after the Upper Triassic and long before the extrusion of the Columbia River basalt. The youngest of the strata cut by these rocks (Pl. 6-A) are thought to be Jurassic, although available evidence does not preclude the possibility that they extend into the Cretaceous. Lindgren[24] concluded that the intrusive activity in the general region accompanied an uplift that took place previous to Chico (Upper Cretaceous)

---

[24] Lindgren, Waldemar, The gold belt of the Blue Mountains of Oregon: U. S. Geol. Survey 22d Ann. Rept., Part 2, p. 596, 1901.

time, because beds of that age on the lower John Day River were not disturbed by the uplift. If any of the younger Mesozoic rocks cut by granodiorite and related rocks in the Wallowa region is Cretaceous, then intrusion must have taken place during the Cretaceous period. It is, however, possible that magmatic activity began during the Jurassic, especially as the diorite-gabbro complex and albite granite do not extend in present exposures into rock younger than Triassic.

APLITE AND RELATED ROCKS

The stocks above described have numerous dikes and other small intrusive masses in and near them. Those of silicic and intermediate compositions are thought to be related to the stocks. Most of them are much too small to show on Plate 1, and only one mass, an oligoclase bostonite stock, has been mapped.

The oligoclase bostonite, a peculiar rock, crops out near the head of Norway Creek. It is a hard, dark bluish gray rock of dense trachytic texture, with scanty small feldspar phenocrysts. Its appearance is strikingly uniform throughout the stock and is composed almost exclusively of oligoclase, with a few small shreds and grains of secondary chlorite and epidote. Dikes of megascopically similar rock are exposed in the valley of the Imnaha River, and an irregular body of it crops out southeast of the confluence of the Cliff and Imnaha Rivers.

The dikes include granitic and granodioritic porphyries, diorite, aplite, and pegmatite. Dark-colored dikes of diabasic and similar compositions are conspicuous in the region. Most of these are believed to be much younger than the dikes here described (pp. 56-57), but some may be of similar age. Goodspeed has postulated[25] that peculiar features in certain dikes near Cornucopia result from interactions between magma and the more or less schistose hornfels that borders and forms xenoliths in the stock in that locality (Pl. 6, *B*). Even in the field, it is clear that the hornfels has been recrystallized and has absorbed

---

[25] Goodspeed, G. E., Effects of inclusions in small porphyry dikes at Cornucopia, Oreg.: Jour. Geology, Vol. 35, No. 7, pp. 653-662, 1927; The mode of origin of a reaction porphyry dike at Cornucopia, Oreg.: Jour. Geology, Vol. 37, No. 2, pp. 158-176, 1929; Some effects of the recrystallization of xenoliths at Cornucopia, Oreg.: Am. Jour. Sci., 5th ser., Vol. 20, pp. 145-150, 1930.

material of igneous origin. The relations are intricate and in many places the contacts between hornfels and igneous rock are gradational.

The aplitic dikes associated with the gold-quartz veins near Cornucopia are of particular interest, because of their relation to the genesis of the deposits. They are mineralized in part and consequently are older than the ore deposits. As Swartley[26] has stated, they are probably among the latest of the dikes connected with the granitic rocks, which they cut. The largest dike is the one that is flanked by branches of the Last Chance vein, but there are several smaller and more irregular dikes of similar composition on the lower levels of the Last Chance, in the Union-Companion and Red Jacket mines, and elsewhere. Most of the dike rock is fine-grained and light-gray, cream, or buff in color, but some is pinkish. The main dike in the Last Chance is made up of allotriomorphic grains whose average diameter is less than 0.05 millimeter in a groundmass of tiny microliths. Some of the smaller dikes are composed of grains about a quarter of a millimeter in average diameter, some of which have a tendency to crystal outlines. Feldspar phenocrysts half a millimeter and more in length are locally present. In many of the rocks the feldspar appears to be albite or albite-oligoclase. Microcline is present in places. Quartz is abundant and in some of the dikes forms about half the rock. Muscovite and, locally, biotite, mostly in small shreds, are scattered throughout but probably do not make up more than 10 per cent of the rock on the average.

The principal quartz veins in the mines at Cornucopia, some of which lie on the borders of aplite dikes, consist of quartz, calcite, and metallic minerals, with muscovite commonly present but generally in minor amount. Apatite and locally epidote are also present and in several places on Two Granites Mountain joint faces in the granitic rock are coated with epidote crystals and, more rarely, with radial aggregates of tourmaline blades. A number of the smaller veins and irregular stringers appear to grade into siliceous dikelets of distinctly igneous appearance and composition. These veins contain, in addition to the quartz, more or less abundant muscovite, locally biotite, and alkali

---

[26] Swartley, A. M., Op. cit., p. 32.

feldspar. In some of the more dikelike veinlets the feldspar is oligoclase. In all the veinlets the feldspar is less abundant than the quartz, and in most it probably does not form as much as 10 per cent of the whole. The dikes associated with these veinlets are composed of feldspar, which is principally either albite or oligoclase, biotite, muscovite, and abundant quartz. In thin section the contacts between material composed dominantly of quartz and that containing much feldspar are sharp. Stringers of quartz penetrate the more typically igneous rock, showing that the quartzose material consolidated later. However, quartzose and feldspathic material occupy parts of the same irregular fissures and differ from each other only in the proportions of the constituents. Their close relationship is, therefore, evident. Locally the veinlets constitute the feather ends of the dikelets.

Cenozoic Stratified Rocks

COLUMBIA RIVER BASALT

Definition and Extent—The name Columbia River basalt is applied to the unmetamorphosed basaltic flows with minor fragmental material that rest on the eroded surface of Mesozoic and older rocks. These flows surround and overlap the Wallowa Mountains and occur as erosion remnants over much of the southeasterly plateau. The formation is considerably more extensive than any of the others.

Character.—The Columbia River basalt in the Wallowa region consists of flows with subordinate layers of tuffs, other pyroclastics, and some sedimentary rocks intercalated locally. The flows vary in thickness from a few feet to 150 feet or even more. The maximum exposed thickness of the series is 2,000 feet. There are more than 20 separate flows in some places.

All of the effusive rocks in this formation appear to be normal basalt. Most are fine-grained, black to dark-brown vesicular rocks; a few are pumiceous and red. In many of the flows the vesicles are partly or entirely filled with zeolites and other minerals; some are not vesicular. The majority are black, cryptocrystalline basalts. A few are jet black, break with a conchoidal fracture, and have a glassy appearance. In most places the basalts are not porphyritic, but in some localities, notably Boulder Creek, there are numerous lath-shaped phenocrysts of labradorite.

As viewed under the microscope, the rocks are principally labradorite and augite. Olivine appears to have been an original constituent locally but in the material studied has been completely altered to serpentine and similar minerals. The other minerals are mostly fresh, although a little chlorite and sericite are present.

Tuff and other clastic rocks are very subordinate to the flows. Thin beds of coarse pyroclastics occur in the southeast portion of the area mapped. There are tuffaceous beds near the mouth of Little Eagle Creek. A considerable area of tuff west of Sparta is described below. Near Richland, farther east, sedimentary rocks, described on pages 62-64, are intercalated in the basalt.

Age.—As has already been stated, the Columbia River basalt of the surrounding region is of Tertiary age and probably in large part Miocene. The evidence at hand indicates that the portion of the formation in the Wallowa region is Miocene, as the only intercalated fossiliferous beds found appear to be of that age (pp. 55-56). The lava is in general not greatly weathered or otherwise altered, but, on the other hand, none of the beds observed appears to be so fresh as to suggest a Quaternary age.

TUFF

Definition and Distribution.—A few miles west of Sparta there are outcrops of tuff which cover a considerable area and form a definite lithologic unit, which has been mapped separately. Only about two square miles of the tuff have been mapped, but similar material, associated with other clastic rocks, extends some distance farther to the west.[27]

Character.—The tuff is a light-gray, porous, and gritty rock, moderately fine-grained as a whole, but with numerous coarse fragments in it. The typical rock is composed principally of glass, with a few admixed crystalline fragments. In many

---

[27] Gilluly, James, Copper deposits near Keating, Oregon: Geol. Survey Bull. 830, pp. 20-21, 1933; Geology and mineral resources of the Baker quadrangle, Oregon: Geol. Survey Bull. 879, pp. 59-63, 1937.

places bedding cannot be discerned, but the rock breaks readily in thin slabs, which are probably parallel to the stratification. In the southern part of the area the tuff is less compact and is somewhat shaly. In this locality it has probably been resorted to some extent by water. In the eastern part of the area of tuff, north of the Baker post road, the tuff locally contains small pebbles of the underlying granite.

Age.—The tuff rests in part on Columbia River basalt, but to the north, east, and west, extends beyond the basalt and rests on granite and older rocks. In the north the tuff is overlain by a considerable thickness of basalt—that is, it is interbedded with the Columbia River basalt near the base of that formation as here exposed. For further discussion see pages 55-56.

CLASTIC ROCKS IN EAGLE VALLEY

Definition and Distribution.—Sedimentary beds intercalated in the Columbia River basalt are found at many places bordering Eagle Valley. The best exposures are in the south bank of Powder River, between Squaw and Daly Creeks. Other good exposures border Immigrant Gulch. Similar beds are intercalated in the basalt near stream level in the canyon of Powder River most of the way between Eagle and Snake Rivers. As these beds largely crop out in bluffs, they are too small to show on the geologic map, Plate 1.

Character.—The formation consists of alternating layers of gravel, sand, clay, and diatomaceous earth, with a few layers of lignitic material. At the largest exposure, that on Powder River near Squaw Creek, the thickness is estimated as about 200 feet, but the base is not exposed. Most of the rocks here are well-bedded, locally cross-bedded cream and yellow sandstone, rather loosely cemented. There are all gradations from very fine sand to small well-rounded pebbles. The beds are broken in several places by slips. There is also coarse gravel, weakly cemented. The cobbles, which are chiefly basalt, are well-rounded, but not well sorted. They are commonly between one and 10 inches in diameter, with some larger and smaller pebbles. Most of the coarse gravel is in the lower part of the exposure. At least two beds of white to yellowish diatomaceous earth occur, the upper about two feet thick, the lower perhaps four feet. The contact of the fine-grained diatomaceous earth with the coarser sediments is strikingly abrupt. The lower of the two beds rests directly on coarse gravel. Impressions of leaf and twig fragments are rather plentiful in these beds. Elsewhere the exposures of the sediments are comparatively poor and small. On and near the old road from Richland to Halfway, on the north side of Eagle Valley, there are diatomaceous beds.

In the southeastern part of Eagle Valley there are one or more beds of impure lignite. The exposure is in a terrace face and is overlain and surrounded by Quaternary gravel. The rock enclosing the lignite is a compacted gray-green clay.

Age.—Leaf impressions found by Mr. Heupgen on the north side of Eagle Valley, in the SW¼ SW¼ sec. 18, T. 9 S., R. 46 E., on the west side of a branch of Immigrant Gulch, have been examined by F. H. Knowlton. He reports: "This species is *Sapindus oregonianus* Knowlton, described originally from the Mascall formation of Van Horn's Ranch, 12 miles west of Mount Vernon, Grant County, Oregon. So far as can be determined from one specimen the age should be the same as the Van Horn's Ranch locality, namely, Mascall". Most of the fragments found were similar in general appearance to the specimen collected.

The material containing these plant impressions was examined by Dr. Albert Mann, of the Carnegie Institute, who states that it is "fresh water diatomaceous earth made up chiefly of one diatom, *Melosira lyrata* Grun, mixed with some microscopic quartz sand and a few fresh-water sponge spicules".

In the nearby Baker quadrangle, Gilluly[28] has mapped extensive deposits like those of Eagle Valley and the more tuffaceous beds near Sparta. In the course of his work, leaves and other vegetable remains were collected in sec. 25, T. 8 S., R. 41 E., and Sec. 10, T. 8 S., R. 42 E. R. W. Brown reports that they are related to the Latah and Mascall flora and are to be regarded as Miocene. Gilluly points out that beds with similar floral content have been found in various relations to basaltic

---

[28] Gilluly, James, Geology and mineral resources of the Baker quadrangle, Oregon: Geol. Survey Bull. 879, pp. 59-63, 1937.

flows in many places in the Northwest, and that most such beds are in restricted bodies, never interconnected. He concludes that, while the beds in the different localities may well be of roughly the same age, correlation such as would be implied by extending such names as "Payette" or "Mascall" to them is not justified. With this conclusion, the present writer is in accord.

Tertiary Dike Rocks

Definition and Distribution.—Beside the metamorphosed dike rocks and those related to the batholithic masses, there is a third group of dikes that consist of unmetamorphosed basalt and are prominent on many mountain sides. The best known examples of such dikes are those on Two Granites Mountain (Pl. 6, *B*) and in Pine Basin near Cornucopia, but similar rocks, although generally of lighter color, were noted on Humming Bird Mountain, along Kettle Creek, on Krag Peak and the contiguous mountain near Cliff River, and in several other places. Because most of the dikes are small and their trends complex, they are not shown on Plate 1. An irregular intrusive mass of trap at the head of the Imnaha was mapped and because of its similarity of appearance, considered to be contemporaneous with these younger dikes.

Character.—These dikes are clean-cut and from a few feet to several rods wide. Those near Cornucopia are fine-grained, black basalt. At least one is vesicular[29] but most are not. Brown diabase dikes are prominent in many places. These cut the Mesozoic and older rocks in various directions, in some places, notably on the south side of Imnaha River above the mouth of Cliff River, making a veritable network. The diabase is unmetamorphosed and fairly fresh.

Age.—Basaltic and diabasic dikes cut all the pre-Tertiary rocks of the region and are unmetamorphosed. Basalt dikes also cut the quartz veins at Cornucopia, in some without appreciable disturbance of the vein; in others with distinct offsets. The brown diabasic dikes so common in the high mountains were not noted at Cornucopia and their relations to the veins are not known, but they so resemble the basalt dikes that they are believed to be related.

---

[29] Dobson, C. G., Private letter of November 17, 1921.

The relations outlined above suggest that the basalt and diabase dikes are of Tertiary age. This combined with the striking similarity of the basalt dikes to the Columbia River basalt flows indicates a genetic connection between the two. The vesicular texture of at least one of the dikes near Cornucopia shows that it was intruded close to the surface under pressure so slight as to permit the escape of gas contained in the cooling magma. There seems little doubt that, as first suggested by Lindgren[30], these dikes were the feeders for the portion of the Columbia River basalt that lies in and around the Wallowa Mountains. Lindgren reports one occurrence in which such a dike appears to be actually connected with one of the Columbia River flows.

Unconsolidated Sediments

Subdivisions and Distribution.—The different deposits of unconsolidated gravel and sand that occur along stream courses in this region are all shown in one color on the geologic map, Plate 1, because without the most detailed physiographic study, it is impossible in many places to distinguish between the different types. These deposits can be placed in at least three categories. These are (1) high level gravel; (2) glacial and fluvio glacial deposits; and (3) flood plain deposits. More thorough examination might permit further subdivision.

On Eagle Creek and some of its tributaries there are benches of gravel and sand rising 50 to 100 feet and more above the present stream level. On East Eagle Creek, and some others, small irregular patches of gravel (not mapped) lie as much as 1,000 feet above the present stream level. They were observed only downstream from the mouth of Hudson and Little Kettle Creeks. Above these, high level gravels appear to be absent along East Eagle Creek, at least for several miles beyond the area mapped. South of Kettle Creek, at altitudes of 6,000 to nearly 7,000 feet, alluvium locally at least 60 feet thick rests on basalt and granitic rock. The upper surface is terraced.

Glacially smoothed and striated rock outcrops are abundant along the Imnaha as far down as the place locally known as the "Blue Holes". High gravel is not known farther upstream,

---

[30] Lindgren, Waldemar, The gold belt of the Blue Mountains of Oregon: Geol. Survey 22d Ann. Rept., Part II, p. 741, 1901.

except in and near some of the tributary gulches, notably on the divide at the head of Blue Creek, where it is abundant 1,200 feet above the river. Probably such high level gravels as may have existed in the valley of the main stream were removed by glacial action, but more thorough examination might reveal small remnants of gravel.

Considerable amounts of gravel occur on Pine Creek and some of its tributaries, both in the stream beds and above them. Nearly all the gravel above the flood plains is on the west side of Pine Creek, where it extends as much as 800 feet higher than the present stream bed. There is coarse gravel in the bed of Boulder Creek, one of the tributaries of Pine Creek, for 1½ miles above its mouth. This gravel is mostly well-rounded and too coarse to have been transported by the present stream. It is therefore thought to be of glacial origin. The valley of Boulder Creek shows no evidence of having been glaciated. Carson Creek also contains much gravel in thick deposits such as those that line the west bank of Pine Creek and it is not confined to the present stream bed.

In the plateau east of Cornucopia, deposits of alluvium outside of the channels and flood plains of existing streams are rare. The only mapped area of older gravel in this part of the region not apparently related to glaciation, is the small patch resting on Columbia River basalt east of Clear Creek at an altitude of about 5,700 feet.

Large gravel deposits occur near Sparta. The thick deposits are rather close to the present streamways, partly because of reconcentration of older deposits; but well-rounded stream gravels are scattered widely over the lava plateau in the general vicinity and particularly to the east of Maiden Gulch. One area has been mapped here. This is partly fine sediments, and partly coarse gravel reconcentrated in a depression in the lava surface. Much of the gravel near Maiden Gulch is too thin and irregularly distributed to be shown on a map of the scale of Plate 1.

Deposits of definitely glacial origin are known only in a few localities, all in the eastern part of the plateau area. The poorly assorted and more or less angular gravel that covers nearly a square mile near the headwaters of East Pine Creek seems clearly morainal. Probably glacio-fluvial gravel occurs farther down this stream and on Clear Creek, Trail Creek, and elsewhere. By far the more extensive and unmistakable glacial deposits found are those west and southeast of Russell Mountain. A morainal ridge nearly three miles long extends southeast from near Fish Lake. South of it is outwash material, for the most part in such thin deposits that it was not mapped. Between the moraine and Duck Lake, three miles to the north, is kettle and kame topography and bedrock is concealed almost everywhere by a covering of till and outwash over an area of some eight or 10 square miles.

The deposits of the present streams are widespread throughout the region. Nearly all the streams are floored with gravel through much of their courses. These deposits, where of sufficient size, are indicated on the map, but the mapping may not be entirely consistent in this respect, as the unconsolidated material was only studied incidentally to the bedrock. Some of the larger occurrences of younger stream deposits are described below.

From Little Kettle Creek down to Sullivan Creek the valley of East Eagle Creek is floored with nearly flat alluvium, largely gravel. The stream is incised in this to a depth of perhaps 10 feet, and has in places cut laterally so as to form a flood plain. Indications of two or more terraces above this were noted. The terracing is particularly well marked between Gold King Creek and a point opposite the mouth of Twin Canyon, on the right side of East Eagle Creek. Glacial striae occur near the mouth of Little Kettle Creek, but no evidence of glaciation was observed below this point.

Alluvial flats occur at the mouths of the major tributaries of the Imnaha River, but chiefly below an altitude of 4,700 feet. No terraces were noted along the upper course of the stream. Below 4,700 feet altitude the amount of alluvial material increases markedly.

Flood plain deposits are present in places along Powder River, but in the region mapped are neither extensive nor thick, and the alluvium in the canyon of Powder River seems less in proportion to the size of the stream than in many of the mountain creeks. Well developed terraces were not observed along Powder River, except in the alluvium of Eagle Valley.

The alluvium that lies in Eagle and Pine Valleys forms the largest deposits of such material in this area. The thickness is uncertain but probably of the order of 200 or 300 feet.

Character.—Except in Pine and Eagle valleys and the parks in the plateau area, the alluvium is nearly all gravel, with only subordinate amounts of sand. Some of the older gravel is sufficiently compacted to stand in bluffs scores of feet high, but has been little cemented.

The placer cuts on Eagle Creek expose distinctly stratified and well-sorted beds of gravel and sand. Most similar deposits on the flanks of other valleys are less perfectly stratified. Both high and low level gravels, here and elsewhere in the area, carry subangular to rounded pebbles and cobbles, such as would be expected in the deposits of a mountain stream. In some localities, such as south of Kettle Creek, the gravel is mixed with yellow clay.

In the definitely glacial deposits in the eastern part of the plateau, nearly every kind of rock that crops out in the Wallowa Mountains can be found, in all size gradation from silt up to boulders the size of a small house. In the moraine east of Fish Lake, particularly, there is great heterogeneity of character and size of the material.

Age.—The unconsolidated sediments in this region range considerably in age. Part of the older alluvium may be preglacial, and, hence, Tertiary. The gravel on the basalt plateau between Sparta and Richland may be late Tertiary, as suggested by Lindgren[31], or it may be early Pleistocene.

The glacial deposits can be confidently referred to the Pleistocene. Although glaciers persist in the Wallowa Mountains, the large ice streams that filled and actively eroded the stream valleys belong to the past.

The younger stream deposits occupying the valley bottoms are all of Quaternary age. The oldest parts may be partly glacial outwash and, hence, Pleistocene, but most are postglacial.

---

[31] Lindgren, Waldemar, The gold belt of the Blue Mountains of Oregon: Geol. Survey 22d Ann. Rept., Part II, p. 747, 1901.

STRUCTURE

The present structure of the rocks of this region is the result of several periods of earth movement. The stresses acting at these different periods differed in direction and intensity, so that the result is complex. The major structural features, so far as they have been deciphered, are described below.

Pre-Tertiary Earth Movements

Granitic rock was intruded into the rocks of this region at some unknown but ancient date. Granitic pebbles were noted in the oldest conglomerates exposed. Doubtless earth movements were associated with this intrusion, but the record of such movements cannot be deciphered. These events took place not later than the Paleozoic era and may have been earlier.

Another period of movement concerning which little information is available is that in which the greenstone and earlier sedimentary rocks were folded. This occurred in the Permian or early Triassic. Notable folding of the strata took place at this time, but the effects of so many later disturbances are superimposed on these old folds that their trends and magnitude have not been ascertained. Pronounced anticlines and synclines with minor crenulations on them were formed. The intensity of folding may have been greater than at any subsequent period.

The orogenic disturbances that produced the structural features now evident in the rocks took place late in the Mesozoic era.

Two interrelated processes were active during this period— folding of the strata under horizontal stress, and granitic intrusion. The folding of the strata started before the stocks began to be intruded into their present positions, and continued until the close of intrusive activity. The thick series of strata were greatly compressed, uplifted, and crumpled into long, narrow anticlines and synclines, with minor crenulations in them.

The inferred trends of the principal folds in the Wallowa region have been approximately indicated in figure 2. Perhaps detailed information would show some of the folds on Eagle

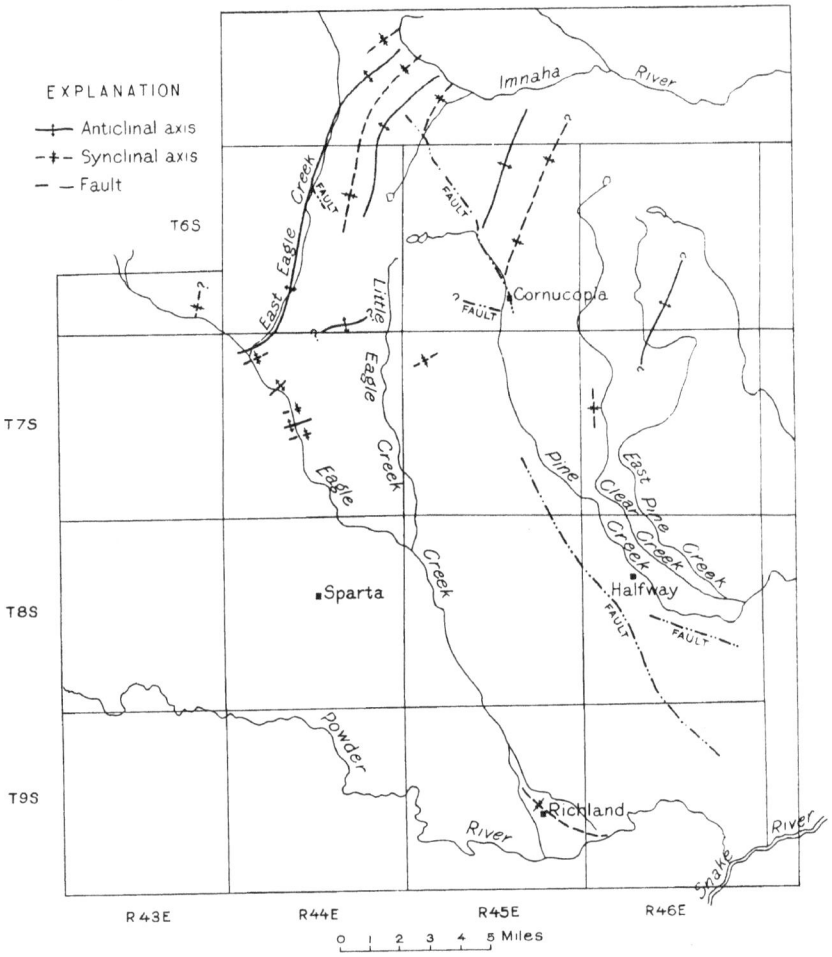

Figure 2—Outline map showing approximate location of Major Tectonic Lines in the Wallowa Region.

Creek to be continuations of the long folds farther north. The broad cover of Columbia River basalt and partial obliteration of bedding by metamorphism obscure the relations. In the high mountains the average trend of the folds is about N. 25° E. This constitutes a conspicuous exception to the conclusion reached by Gilluly and his associates that in this part of Oregon the folds in the pre-Tertiary rocks strike nearly east[32]. South of the mountains, along Eagle Creek, the folds swing more to the east and some trend due east. In the plateau country to the east of the mountains the structure is largely obscured by a cover of basalt and unconsolidated deposits. The average trend here seems to be only a little east of north, but locally strikes are well to the east of north, and in places almost due east. The structure is complicated because the rocks had all been previously folded before the present folds were begun. Since little definite information is at hand regarding the ancient folding, there is no way of estimating its effect on the form of the later folds.

The character of the folds in the high mountains is illustrated in the western parts of structure sections AA^1 and BB^1 on Plate 1. The broader folds in the plateau area are shown in the eastern parts of these sections and in sections CC^1, DD^1 and EE^1 on the same plate. The irregular lower contact of the Columbia River basalt is indicated on all these sections except DD^1. The pronounced fault in West Basin is illustrated in section CC^1.

The folding in the high mountains is more intense and complex than elsewhere. Here the maximum uplift occurred and granitic intrusion was most active. The folds are tightly compressed and somewhat overturned. There is evidence of considerable minor crumpling within the major folds. The details of these minor crenulations await further work for their elucidation. The relative spacing of the folds in the high mountains and the plateau is shown in figure 2. The tectonic lines shown in this figure are based on interpretation of the field relations of the rocks as shown on Plate 1, and no great accuracy is claimed for them. Like the structure sections in Plate 1, they are diagrammatic.

---

[32] Gilluly, James, Reed, J. C., and Park, C. F., Jr., Some mining districts of eastern Oregon: Geol. Survey Bull. 846, p. 20, 1933.

The folds in the high mountains are peculiar in that some are overturned in opposite directions to others. The anticline on East Eagle Creek is overturned to the west and closely associated folds on Imnaha River, only a short distance to the north, are overturned to the east. This is suggested in section AA, Plate 1. The overturning is more pronounced north of the Imnaha, just outside the region mapped. The glaciated exposures here show that the syncline east of the anticline of East Eagle Creek is quite as sharply overturned to the east as the anticline farther south is overturned to the west. It is as though the group of folds were pinched in below and flared out above, somewhat after the manner of a fan. This group of folds is underlain by granodiorite and related rocks, and included between two great masses of such rock. It may be that the peculiar arrangement of these folds is related in some way to the intrusion. The already folded strata might be conceived to have been caught between the two great prongs of the intruding magma bodies, so that the lower parts of the folds were pinched together, causing them to flare out above.

Overturning is not confined to the high mountains. A tightly folded syncline overturned to the north is exposed by placer operations on Eagle Creek below Martin Bridge. This is illustrated in section DD', Plate 1. This syncline is just north of a felsitic intrusive connected with the albite granite stock. There is probably some genetic relation between the fold and the intrusion. If the intrusion preceded the overturning, the latter may be the result of thrust against the igneous massif or possibly the force of intrusion compressed and overturned pre-existing fold. Probably the actual process was some combination of these two.

The granitic intrusions not only truncated the strata, but also bent and twisted them out of their path. Most of the folding in the region certainly preceded the intrusion as the granitic rocks show no evidence of having been subjected to the stresses that caused the folding. Comparatively minor folding, however, accompanied the intrusion. The elliptical mass of granodiorite and similar rock near Cornucopia is well exposed, and its relations to the surrounding rocks can be clearly made out in several places. The contacts are everywhere steep—where measured they dip 70° or more. The intrusion has truncated the older strata, and in places shattered them. There are large included blocks of sedimentary rock in the granodiorite near Cornucopia. On Pine Creek, above the Queen of the West mill, the contact is highly irregular, with tongues of granodiorite penetrating along cracks in the shattered hornfels and fragments of hornfels included in the igneous mass. Irregular dikes of granodiorite and aplite are abundant in the slate for some distance from the contact. (See Plate 6.) Blocks of hornfels were probably assimilated by the magma and material from the magma has soaked into and partially changed the hornfels that remains. Extensive shattering of the older rock and intrusion of dikes in the cracks thus created occurred about the small granitic mass on Corral Creek and, less obviously, at other places. Bowing up of the older strata parallel to the contact is strikingly shown on both the east and west sides of Granite Mountain. The upbending is abrupt and does not extend far out from the contact. It seems that the strata were forced up by the rising magma. This shattering and bending of the strata would seem to indicate that the rocks were under no great load and were, therefore, not buried deeply below the surface at the time of the intrusion. However, little diminution in the grain size of the granodiorite was observed at the contacts.

The larger granitic mass of Humming Bird Mountain was intruded in a similar manner. Here also older sediments are engulfed in the granite in the southeast part of the mass. Exposed contacts are few and confined to the northern part but, so far as can be judged, they are steep. The overturned fold in the limestone below Martin Bridge suggests that the intrusion bent the strata round it.

The structure of the rocks underlying the eastern part of the plateau area is less definitely known than that of any other part of the region where pre-Tertiary strata are exposed. This is partly because the bedding has been obscured by metamorphism and, hence, the stratigraphic relations are not understood and satisfactory outcrops are scarce. The dominant structural texture here seems to be a broad anticline whose trend is not certainly known. It is indicated, tentatively on figure 2, as striking almost due north, but actually may trend much more to the east. To explain the apparent relations of this anticline to the structure in the high mountains, one of two things must be assumed. Either there is a great fault passing through the area occupied by the diorite-gabbro complex, or else the sedimentary rocks east of the intrusion are older than the oldest recognized west of it and underlie rather than rest on the greenstone. The apparent discrepancy in the structure, which must be explained by such hypotheses as these, is clearly brought out in section AA[1] on Plate 1, and discussed below.

A syncline, indicated on figure 2, follows the lower course of Clear Creek, but east and north of the two intrusive masses the strata dip consistently westward until near the place where they are overlapped by the great sheet of Columbia River basalt, which extends from here to Snake River. A short distance before they disappear under the lava, eastward dips are noted. No repetition by faulting or folding was detected, but the stratigraphic succession in this thick series is not well known, and duplication may be present without having been recognized. However, any duplication that may exist is not believed sufficient to account for the apparent difference in thickness between the rocks in this part of the region and in the high mountains.

No evidence of notable faulting, such as is required in the first hypothesis suggested, was observed. The work was necessarily somewhat hasty and the outcrops were not everywhere satisfactory, so that minor faults probably escaped discovery. A fault necessary to satisfy the hypothesis would have a throw of thousands of feet and extend for miles. So great a fault would surely have left conspicuous shear zones and other evidence of its existence. The simplest explanation that fits the known facts is the second hypothesis suggested—that is, that the structure here is in general a broad arch in which sedimentary rocks, in part or whole more ancient than any recognized elsewhere in the area, are exposed east of the diorite-gabbro complex. It may well be that this general structure is complicated by minor folds and faults not yet deciphered.

Several faults are known in the pre-Tertiary rocks, and others are suspected. Only the larger faults observed could be shown on the map, Plate 1. Other smaller faults were noted, and probably some as extensive as those mapped have escaped observation. In general, however, faulting in the pre-Tertiary rocks is subordinate to folding.

THE WALLOWA MOUNTAINS

The largest fault known is that which extends from near Cornucopia through West Basin and across the Cliff River Valley. It is best exposed in West Basin, and may for convenience be termed the West Basin fault. The trace of the fault is distinctly curved, but in West Basin has an average strike of about N. 30° W. The dip, as determined graphically, from its intersection with the surface as shown on Plate 1, is about 55° SW. The downthrow is to the southwest and of the order of several thousand feet. A sliver of greenstone, evidently wedged up on the line of the fault, crops out in the gap in the ridge above West Basin. The fault evidently continues across Cliff River, as indicated in Plate 1, thus accounting for the absence of the Martin Bridge formation, so prominent along East Eagle Creek and near the head of Imnaha River. Poor outcrops and the contact metamorphism around the granitic mass near Corral Creek have added to the difficulties of the mapping, which is here generalized; these rocks have been mapped as younger Mesozoic strata, but Triassic (?) volcanic rocks may also be present west of Cliff River. Near the place where a vertical dip symbol is drawn about a mile north of the stock black slate is exposed. This has not been differentiated in mapping from the Carboniferous (?) rocks around it.

The fault system in the valley of Cliff River is more complex than could be indicated on the map. The single fault line trending nearly west shown on the ridge above Clipper Basin represents a group of small faults, well exposed. The distribution of the rocks near Cliff River indicates the presence of faults but they could not be traced in detail and are all subsidiary to the West Basin fault. Another subsidiary fault is suggested by the distribution of the formations in Elk Creek Valley, below Cornucopia, near the end of the West Basin fault, but the abundant gravel conceals the structure.

A fault was observed in the valley of Little Kettle Creek and similar ones probably occur along East Eagle Creek and its tributaries, as indicated on Plate 1. The absence of the Martin Bridge formation on Lime Creek and the angularity of the limestone block to the south suggest that the latter may be bounded on the northeast by a fault. Minor faults were observed by Jacques Heupgen near the Sheep Rock mine and the northeast side of the block of limestone here may also be bounded by faults. These faults appear to strike N. 30°-45° W. The one on Little Kettle Creek appears to be downthrown on the northeast, and the others presumably on the southwest. Dips were not observed and the contours on Plate 1 are too generalized to permit of inferences as to dip. Some small faults were noted near Martin Bridge, the largest on Eagle Creek, less than half a mile above the bridge. It strikes northeast and has a downthrow of about 40 feet to the northwest.

Structures in the Tertiary Rocks

Probably most of the Columbia River lava in the Wallowa region rose through the fissures now filled with basaltic dikes in the high mountains. In several places dikes cut the flows. These are particularly prominent near Robinette, on Snake River. Such dikes doubtless contributed to the lava mass, which eventually covered the country surrounding the mountains to depths of as much as 2,000 feet.

The rise of so much lava evidently disturbed the equilibrium of the earth's crust. A general tilt away from the mountains was produced, apparently sufficient to produce a difference in elevation of about 2,000 feet in a distance of 18 miles. Various minor flexures and fractures were also produced. The most prominent are those that resulted in the depressions now occupied by Pine and Eagle Valleys.

Pine Valley coincides with a downfaulted block or graben in the Columbia River basalt. The major faults on the south side of the valley can be easily seen. Their positions have been roughly indicated on figure 2. The variable dips of the flows and irregular topographic forms of the basalt blocks near Sunset, in the east part of the valley, strongly suggest faulting but the positions of the faults have not been determined. On the north side of Pine Valley the flows dip towards it and may pass under the alluvium without any breaks.

Eagle Valley is a synclinal trough. The basalt flows dip on both sides towards it but no faults were noted. The axis of the syncline is roughly parallel to the elongation of the valley, as indicated on figure 2. At the southeast end of the valley the syncline seems to swing around to a position approximately coincident with the valley of Powder River. The dip of the flows here, however, varies so much that the dominating structural feature was not definitely ascertained.

Many of the faults in the older rocks are approximately parallel to the major faults and folds in the Columbia River basalt. The strike varies but averages about N. 30° W., in contrast to the northeastward trend of the principal folds in the older rocks. Direct evidence as to the relation of the Tertiary rocks to the faults in the older strata was not obtained. The lack of topographic expression suggests that no appreciable movements have recently taken place along the faults in the older rocks. The small stock on Cliff River is near the intersection of several faults and may have been intruded into the zone of weakness resulting from the faulting. If so, the faults must antedate the intrusion, which is probably not younger than Cretaceous. Evidence in support of this view is afforded by the veins near Cornucopia, which are related to the granitic stock there. Most of the principal veins are on the west, or hanging wall, side of the fault that extends from Pine Creek across West Basin and in the angle between that fault and that which probably exists along Elk Creek, at the southeast end of the stock. These veins converge toward a point on or near the fault along Pine Creek. The vein fissures are thought to be subsidiary fractures related to the main faulting, which implies that the latter are of pre-Tertiary age. The faults and related folds in the Columbia River basalt may, however, lie along preexisting lines of weakness and indicate renewed movement along old fault lines.

GEOLOGIC HISTORY

Pre-Carboniferous Time

In the Wallowa region no rocks older than Carboniferous have been recognized, so that the earlier history can be surmised only by comparison with other regions. Almost nothing is known regarding the pre-Cambrian in and near the Wallowa Mountains. Lindgren[33] found some biotite-gneiss of sedimentary origin in the Elkhorn Range, which he considered of Archean

---
[33] Lindgren, Waldemar, Op. cit., p. 577.

age. This is the only record of supposed pre-Cambrian rocks near the Wallowa Mountains. The Paleozoic record up to Carboniferous time is even more scanty. Although most of the region has been studied in reconnaissance by Lindgren and several other geologists, no Paleozoic rocks older than Carboniferous have been found, and the inference is strong that no such rocks are exposed. With the deep erosion known to have occurred, it is strange, if any great thickness of such rocks exists, that they have remained covered everywhere. Possibly this region was dry land throughout lower and middle Paleozoic and perhaps also during the late pre-Cambrian, that is, it formed a portion of the ancient land mass that has been called "Cascadia".[34]

In the absence of a sedimentary record, nothing can be definitely stated as to orogenic movement during this long period but that such movements took place is probable because pebbles of granitoid, mostly gneissic rocks occur in conglomerate of the oldest formation in the area. Lindgren[35] identified quartz diorite pebbles in a conglomerate or breccia associated with the limestone on East Eagle Creek. Pebbles of gneissic granitoid rocks were noted in conglomerate in the rocks underlying the Martin Bridge formation on Spring Creek and elsewhere. These rocks are clearly far older than the granodiorite and related rocks that are at present prominent in the region. They may have come from some ancient batholith of Paleozoic or even earlier date.

Carboniferous Time

Much remains to be learned regarding the events of Carboniferous time in this region, but some important facts are already available. A great thickness of strata of this age was deposited here. The paleontologic data, though scanty, suggest that the rocks as a whole were formed in late upper Carboniferous time, and the only adequate collections of fossils indicate Permian age. The Carboniferous must have been a time of many changes and fluctuations in this region. The sedimentary rocks include conglomerates with poorly rounded and sorted pebbles and cross-bedded sandstone. Such beds were probably

---
[34] Grabau, A. W., A textbook of geology, Part II, p. 214, 1921.
[35] Lindgren, Waldemar, Op. cit., p. 735.

formed in deltas and other near-shore deposits and some may be deposits of streams on the land surface of that time. The limestone and chert, however, are marine and may have been formed at comparatively great depth under the sea. The black slate, perhaps the oldest formation exposed, may also have been formed at a moderate depth beneath the sea.

The great thickness of Permian volcanics gives striking evidence of volcanism on a scale comparable with that of the famous Columbia River basalt. The Permian deposits have not been proved so extensive areally, but their thickness is much greater and their known extent very great. Unlike that of the Tertiary basalt, the effusion of these Permian andesitic flows was accompanied by notable explosive activity and the eruption of great quantities of fragmental ejecta. The eruptions evidently took place near the sea, as their deposits are associated with marine sediments. Some of them may well have been submarine.

The oscillations of level and igneous activity recorded in the Carboniferous rocks were preliminary to greater movements. At the close of the Paleozoic there was great orogenic movement. The rocks were folded and sheared. Probably most of the metamorphism of the Paleozoic rocks occurred at this time. No definite evidence of batholithic intrusion at this time is available but the amphibolite masses on Clear Creek may have been, and some of the metamorphosed dike rocks probably were injected at this time. The apparent intensity of the folding suggests that there was uplift and mountain-making ancestral to the present Wallowa Mountains and neighboring ranges. This theory is strengthened by the apparent absence of Lower and Middle Triassic strata and the evidence of erosion during those epochs.

Mesozoic Time

Most or all of the Middle or Lower Triassic was probably an erosion interval. The absence of strata of that age is not absolutely certain, but the unconformity at the base of the Upper Triassic Martin Bridge formation indicates a considerable time gap during which the older rocks were uplifted and eroded.

In the Upper Triassic the region was again covered by the sea. The limestone and limy shale of the Martin Bridge formation were deposited on the eroded surface of the older rocks. The great thickness of this formation shows that subsidence was long continued. There was clearly some volcanism at this time, and locally the eruptions of lava, ash, and other pyroclastics were probably voluminous. The source of this material is not known, but from their intimate associations with marine deposits, it must have been either near the shore, perhaps on islands, or even submarine. Some of the volcanic deposits near Cliff River seem almost certainly of submarine origin.

No stratigraphic break has been detected between the Martin Bridge formation and the overlying rocks. Deposition of limestone ceased, but marine conditions probably persisted to a date as yet undetermined—perhaps Jurassic or even Lower Cretaceous.

The greatest orogenic movements of which a clear record remains took place in late Mesozoic, probably Cretaceous, time. The batholithic masses of granodiorite and related rocks were intruded. The older rocks were violently compressed and thrown into great folds, metamorphosed, sheared, and probably faulted. Mineral deposits of various kinds were formed in them. The whole Blue Mountain region was uplifted thousands of feet above the sea, which formerly had covered it. There is some evidence that at least as early as Chico (Upper Cretaceous) time the region was a land mass. The known marine deposits of that time do not extend farther east than the valley of the John Day River.

As soon as the region projected above the sea it was attacked by the forces of erosion. Streams cut channels into the surface and began to carry detritus back into the sea from which the land had recently emerged. Uplift proceeded faster than erosion, and the amount of land above the sea continued to increase. As erosion continued valleys were developed between the masses of greatest uplift. The latter were themselves carved by numerous streams. The mountain ranges born at this time have persisted, with various modifications, to the present day.

Tertiary Time

The erosion that started in the late Mesozoic probably continued with interruptions into the early Tertiary. Much of the country was reduced to low relief, but the Wallowa Mountains and similar groups remained above this surface. About this time probably in the late Eocene, came a renewal of uplift in which valleys were cut into the plateau surface around the mountains. This rejuvenation of erosive activity could only have come as a result of relative uplift of the mountains. As it coincides in time with the inception of the volcanic activity, which constitutes the most striking feature of the Tertiary history, there may be a genetic relation as well. Perhaps the cracks in the earth's crust through which the lava floods poured were produced as a result of this movement.

However they started, the eruptions of the Columbia River lava are most impressive. Commencing apparently in the Eocene, they swelled to greatest volume in the Miocene, and may have continued into the Pliocene. The flows in the Wallowa region represent but a small portion of the great mass of lava erupted during this time from fissures such as those near Cornucopia and are thought to be all of Miocene age.

Relative, perhaps absolute, elevation of the mountains and subsidence of the surrounding country, which began in the Eocene, continued throughout Tertiary time. The extent of tilting of the Columbia River basalt indicates relative change in elevation since the start of the eruptions as of the order of 1,500 feet or more. Numerous and profound drainage changes resulted from these earth movements and the eruptions that accompanied the earlier ones. At the end of the Tertiary the major physical features of the regions, as we now know them, had been developed. The Wallowa Mountains, although unglaciated and less rugged than they now are, occupied their present position, surrounded by a plateau already deeply dissected by erosion. Large quantities of the basalt had already been swept away by the streams. Probably the present drainage pattern had in large part been established, although readjustment may have continued into the Pleistocene.

Quaternary Time

If we consider that the Pleistocene epoch began with the coming of glaciation, the geologic work accomplished since the close of the Tertiary has been small compared with that effected previously. During Pleistocene time, glaciers filled the mountain valleys and capped part of the plateau country. They denuded the summits of soil and carved them into the jagged assemblage of pinnacles and cliffs that exist today. The lakes scattered throughout the mountains were produced by the glaciers. The rugged beauty and picturesqueness of the present-day scenery is largely due to glaciation. Yet, striking as the work of the glaciers is, it is trivial compared to the denudation accomplished by the pre-Quaternary streams, which removed thousands of feet of hard strata and cut deeply into the stocks below.

www.GoldRushBooks.com

More Books On Mining

Visit: www.goldrushbooks.com to order your copies or ask your favorite book seller to offer them.

Mining Books by Kerby Jackson

Gold Dust: Stories From Oregon's Mining Years

Oregon mining historian and prospector, Kerby Jackson, brings you a treasure trove of seventeen stories on Southern Oregon's rich history of gold prospecting, the prospectors and their discoveries, and the breathtaking areas they settled in and made homes. **5" X 8", 98 ppgs. Retail Price: $11.99**

The Golden Trail: More Stories From Oregon's Mining Years

In his follow-up to "Gold Dust: Stories of Oregon's Mining Years", this time around, Jackson brings us twelve tales from Oregon's Gold Rush, including the story about the first gold strike on Canyon Creek in Grant County, about the old timers who found gold by the pail full at the Victor Mine near Galice, how Iradel Bray discovered a rich ledge of gold on the Coquille River during the height of the Rogue River War, a tale of two elderly miners on the hunt for a lost mine in the Cascade Mountains, details about the discovery of the famous Armstrong Nugget and others. **5" X 8", 70 ppgs. Retail Price: $10.99**

Oregon Mining Books

Geology and Mineral Resources of Josephine County, Oregon

Unavailable since the 1970's, this important publication was originally compiled by the Oregon Department of Geology and Mineral Industries and includes important details on the economic geology and mineral resources of this important mining area in South Western Oregon. Included are notes on the history, geology and development of important mines, as well as insights into the mining of gold, copper, nickel, limestone, chromium and other minerals found in large quantities in Josephine County, Oregon. **8.5" X 11", 54 ppgs. Retail Price: $9.99**

Mines and Prospects of the Mount Reuben Mining District

Unavailable since 1947, this important publication was originally compiled by geologist Elton Youngberg of the Oregon Department of Geology and Mineral Industries and includes detailed descriptions, histories and the geology of the Mount Reuben Mining District in Josephine County, Oregon. Included are notes on the history, geology, development and assay statistics, as well as underground maps of all the major mines and prospects in the vicinity of this much neglected mining district. **8.5" X 11", 48 ppgs. Retail Price: $9.99**

The Granite Mining District

Notes on the history, geology and development of important mines in the well known Granite Mining District which is located in Grant County, Oregon. Some of the mines discussed include the Ajax, Blue Ribbon, Buffalo, Continental, Cougar-Independence, Magnolia, New York, Standard and the Tillicum. Also included are many rare maps pertaining to the mines in the area. **8.5" X 11", 48 ppgs. Retail Price: $9.99**

Ore Deposits of the Takilma and Waldo Mining Districts of Josephine County, Oregon

The Waldo and Takilma mining districts are most notable for the fact that the earliest large scale mining of placer gold and copper in Oregon took place in these two areas. Included are details about some of the earliest large gold mines in the state such as the Llano de Oro, High Gravel, Cameron, Platerica, Deep Gravel and others, as well as copper mines such as the famous Queen of Bronze mine, the Waldo, Lily and Cowboy mines. This volume also includes six maps and 20 original illustrations. **8.5" X 11", 74 ppgs. Retail Price: $9.99**

Metal Mines of Douglas, Coos and Curry Counties, Oregon

Oregon mining historian Kerby Jackson introduces us to a classic work on Oregon's mining history in this important re-issue of Bulletin 14C Volume 1, otherwise known as the Douglas, Coos & Curry Counties, Oregon Metal Mines Handbook. Unavailable since 1940, this important publication was originally compiled by the Oregon Department of Geology and Mineral Industries includes detailed descriptions, histories and the geology of over 250 metallic mineral mines and prospects in this rugged area of South West Oregon. **8.5" X 11", 158 ppgs. Retail Price: $19.99**

Metal Mines of Jackson County, Oregon

Unavailable since 1943, this important publication was originally compiled by the Oregon Department of Geology and Mineral Industries includes detailed descriptions, histories and the geology of over 450 metallic mineral mines and prospects in Jackson County, Oregon. Included are such famous gold mining areas as Gold Hill, Jacksonville, Sterling and the Upper Applegate. **8.5" X 11", 220 ppgs. Retail Price: $24.99**

Metal Mines of Josephine County, Oregon

Oregon mining historian Kerby Jackson introduces us to a classic work on Oregon's mining history in this important re-issue of Bulletin 14C, otherwise known as the Josephine County, Oregon Metal Mines Handbook. Unavailable since 1952, this important publication was originally compiled by the Oregon Department of Geology and Mineral Industries includes detailed descriptions, histories and the geology of over 500 metallic mineral mines and prospects in Josephine County, Oregon. **8.5" X 11", 250 ppgs. Retail Price: $24.99**

Metal Mines of North East Oregon

Oregon mining historian Kerby Jackson introduces us to a classic work on Oregon's mining history in this important re-issue of Bulletin 14A and 14B, otherwise known as the North East Oregon Metal Mines Handbook. Unavailable since 1941, this important publication was originally compiled by the Oregon Department of Geology and Mineral Industries and includes detailed descriptions, histories and the geology of over 750 metallic mineral mines and prospects in North Eastern Oregon. **8.5" X 11", 310 ppgs. Retail Price: $29.99**

Metal Mines of North West Oregon

Oregon mining historian Kerby Jackson introduces us to a classic work on Oregon's mining history in this important re-issue of Bulletin 14D, otherwise known as the North West Oregon Metal Mines Handbook. Unavailable since 1951, this important publication was originally compiled by the Oregon Department of Geology and Mineral Industries and includes detailed descriptions, histories and the geology of over 250 metallic mineral mines and prospects in North Western Oregon. **8.5" X 11", 182 ppgs. Retail Price: $19.99**

Mines and Prospects of Oregon

Mining historian Kerby Jackson introduces us to a classic mining work by the Oregon Bureau of Mines in this important re-issue of The Handbook of Mines and Prospects of Oregon. Unavailable since 1916, this publication includes important insights into hundreds of gold, silver, copper, coal, limestone and other mines that operated in the State of Oregon around the turn of the 19th Century. Included are not only geological details on early mines throughout Oregon, but also insights into their history, production, locations and in some cases, also included are rare maps of their underground workings. **8.5" X 11", 314 ppgs. Retail Price: $24.99**

Lode Gold of the Klamath Mountains of Northern California and South West Oregon

(See California Mining Books)

Mineral Resources of South West Oregon

Unavailable since 1914, this publication includes important insights into dozens of mines that once operated in South West Oregon, including the famous gold fields of Josephine and Jackson Counties, as well as the Coal Mines of Coos County. Included are not only geological details on early mines throughout South West Oregon, but also insights into their history, production and locations. **8.5" X 11", 154 ppgs. Retail Price: $11.99**

Chromite Mining in The Klamath Mountains of California and Oregon

(See California Mining Books)

Southern Oregon Mineral Wealth

Unavailable since 1904, this rare publication provides a unique snapshot into the mines that were operating in the area at the time. Included are not only geological details on early mines throughout South West Oregon, but also insights into their history, production and locations. Some of the mining areas include Grave Creek, Greenback, Wolf Creek, Jump Off Joe Creek, Granite Hill, Galice, Mount Reuben, Gold Hill, Galls Creek, Kane Creek, Sardine Creek, Birdseye Creek, Evans Creek, Foots Creek, Jacksonville, Ashland, the Applegate River, Waldo, Kerby and the Illinois River, Althouse and Sucker Creek, as well as insights into local copper mining and other topics. **8.5" X 11", 64 ppgs. Retail Price: $8.99**

Geology and Ore Deposits of the Takilma and Waldo Mining Districts

Unavailable since the 1933, this publication was originally compiled by the United States Geological Survey and includes details on gold and copper mining in the Takilma and Waldo Districts of Josephine County, Oregon. The Waldo and Takilma mining districts are most notable for the fact that the earliest large scale mining of placer gold and copper in Oregon took place in these two areas. Included in this report are details about some of the earliest large gold mines in the state such as the Llano de Oro, High Gravel, Cameron, Platerica, Deep Gravel and others, as well as copper mines such as the famous Queen of Bronze mine, the Waldo, Lily and Cowboy mines. In addition to geological examinations, insights are also provided into the production, day to day operations and early histories of these mines, as well as calculations of known mineral reserves in the area. This volume also includes six maps and 20 original illustrations. **8.5" X 11", 74 ppgs. Retail Price: $9.99**

Gold Mines of Oregon

Oregon mining historian Kerby Jackson introduces us to a classic work on Oregon's mining history in this important re-issue of Bulletin 61, otherwise known as "Gold and Silver In Oregon". Unavailable since 1968, this important publication was originally compiled by geologists Howard C. Brooks and Len Ramp of the Oregon Department of Geology and Mineral Industries and includes detailed descriptions, histories and the geology of over 450 gold mines Oregon. Included are notes on the history, geology and gold production statistics of all the major mining areas in Oregon including the Klamath Mountains, the Blue Mountains and the North Cascades. While gold is where you find it, as every miner knows, the path to success is to prospect for gold where it was previously found. **8.5" X 11", 344 ppgs. Retail Price: $24.99**

Mines and Mineral Resources of Curry County Oregon

Originally published in 1916, this important publication on Oregon Mining has not been available for nearly a century. Included are rare insights into the history, production and locations of dozens of gold mines in Curry County, Oregon, as well as detailed information on important Oregon mining districts in that area such as those at Agness, Bald Face Creek, Mule Creek, Boulder Creek, China Diggings, Collier Creek, Elk River, Gold Beach, Rock Creek, Sixes River and elsewhere. Particular attention is especially paid to the famous beach gold deposits of this portion of the Oregon Coast. **8.5" X 11", 140 ppgs. Retail Price: $11.99**

Chromite Mining in South West Oregon

Mining historian Kerby Jackson introduces us to a classic mining work in this important re-issue of the Oregon Department of Geology and Mineral Industries publication "Chromite in South West Oregon". Originally published in 1961, this important publication on Oregon Mining has not been available for nearly a century. Included are rare insights into the history, production and locations of nearly 300 chromite mines in South Western Oregon. **8.5" X 11", 184 ppgs. Retail Price: $14.99**

Mineral Resources of Douglas County Oregon

Mining historian Kerby Jackson introduces us to a classic mining work in this important re-issue of the Oregon Department of Geology and Mineral Industries publication "Geology and Mineral Resources of Douglas County, Oregon". Originally published in 1972, this important publication on Oregon Mining has not been available for nearly forty years. Included are rare insights into the geology, history, production and locations of numerous gold mines and other mining properties in Douglas County, Oregon. **8.5" X 11", 124 ppgs. Retail Price: $11.99**

Idaho Mining Books

Gold in Idaho

Unavailable since the 1940's, this publication was originally compiled by the Idaho Bureau of Mines and includes details on gold mining in Idaho. Included is not only raw data on gold production in Idaho, but also valuable insight into where gold may be found in Idaho, as well as practical information on the gold bearing rocks and other geological features that will assist those looking for placer and lode gold in the State of Idaho. This volume also includes thirteen gold maps that greatly enhance the practical usability of the information contained in this small book detailing where to find gold in Idaho. **8.5" X 11", 72 ppgs. Retail Price: $9.99**

Geology of the Couer D'Alene Mining District of Idaho

Unavailable since 1961, this publication was originally compiled by the Idaho Bureau of Mines and Geology and includes details on the mining of gold, silver and other minerals in the famous Coeur D'Alene Mining District in Northern Idaho. Included are details on the early history of the Coeur D'Alene Mining District, local tectonic settings, ore deposit features, information on the mineral belts of the Osburn Fault, as well as detailed information on the famous Bunker Hill Mine, the Dayrock Mine, Galena Mine, Lucky Friday Mine and the infamous Sunshine Mine. This volume also includes sixteen hard to find maps. **8.5" X 11", 70 ppgs. Retail Price: $9.99**

Utah Mining Books

Fluorite in Utah

Unavailable since 1954, this publication was originally compiled by the USGS, State of Utah and U.S. Atomic Energy Commission and details the mining of fluorspar, also known as fluorite in the State of Utah. Included are details on the geology and history of fluorspar (fluorite) mining in Utah, including details on where this unique gem mineral may be found in the State of Utah. **8.5" X 11", 60 ppgs. Retail Price: $8.99**

California Mining Books

The Tertiary Gravels of the Sierra Nevada of California

Mining historian Kerby Jackson introduces us to a classic mining work by Waldemar Lindgren in this important re-issue of The Tertiary Gravels of the Sierra Nevada of California. Unavailable since 1911, this publication includes details on the gold bearing ancient river channels of the famous Sierra Nevada region of California. **8.5" X 11", 282 ppgs. Retail Price: $19.99**

The Mother Lode Mining Region of California

Unavailable since 1900, this publication includes details on the gold mines of California's famous Mother Lode gold mining area. Included are details on the geology, history and important gold mines of the region, as well as insights into historic mining methods, mine timbering, mining machinery, mining bell signals and other details on how these mines operated. Also included are insights into the gold mines of the California Mother Lode that were in operation during the first sixty years of California's mining history. **8.5" X 11", 176 ppgs. Retail Price: $14.99**

Lode Gold of the Klamath Mountains of Northern California and South West Oregon

Unavailable since 1971, this publication was originally compiled by Preston E. Hotz and includes details on the lode mining districts of Oregon and California's Klamath Mountains. Included are details on the geology, history and important lode mines of the French Gulch, Deadwood, Whiskeytown, Shasta, Redding, Muletown, South Fork, Old Diggings, Dog Creek (Delta), Bully Choop (Indian Creek), Harrison Gulch, Hayfork, Minersville, Trinity Center, Canyon Creek, East Fork, New River, Denny, Liberty (Black Bear), Cecilville, Callahan, Yreka, Fort Jones and Happy Camp mining districts in California, as well as the Ashland, Rogue River, Applegate, Illinois River, Takilma, Greenback, Galice, Silver Peak, Myrtle Creek and Mule Creek districts of South Western Oregon. Also included are insights into the mineralization and other characteristics of this important mining region. **8.5" X 11", 100 ppgs. Retail Price: $10.99**

Mines and Mineral Resources of Shasta County, Siskiyou County, Trinity County, California

Unavailable since 1915, this publication was originally compiled by the California State Mining Bureau and includes details on the gold mines of this area of Northern California. Also included are insights into the mineralization and other characteristics of this important mining region, as well as the location of historic gold mines. **8.5" X 11", 204 ppgs. Retail Price: $19.99**

Geology of the Yreka Quadrangle, Siskiyou County, California

Unavailable since 1977, this publication was originally compiled by Preston E. Hotz and includes details on the geology of the Yreka Quadrangle of Siskiyou County, California. Also included are insights into the mineralization and other characteristics of this important mining region. **8.5" X 11", 78 ppgs. Retail Price: $7.99**

Mines of San Diego and Imperial Counties, California

Originally published in 1914, this important publication on California Mining has not been available for a century. This publication includes important information on the early gold mines of San Diego and Imperial County, which were some of the first gold fields mined in California by early Spanish and Mexican miners before the 49ers came on the scene. Included are not only details on early mining methods in the area, production statistics and geological information, but also the location of the early gold mines that helped make California "The Golden State". Also included are details on the mining of other minerals such as silver, lead, zinc, manganese, tungsten, vanadium, asbestos, barite, borax, cement, clay, dolomite, fluospar, gem stones, graphite, marble, salines, petroleum, stronium, talc and others. **8.5" X 11", 116 ppgs. Retail Price: $12.99**

Mines of Sierra County, California

Unavailable since 1920, this publication was originally compiled by the California State Mining Bureau and includes details on the gold mines of Sierra County, California. Also included are insights into the mineralization and other characteristics of this important mining region, as well as the location of historic gold mines. **8.5" X 11", 156 ppgs. Retail Price: $19.99**

Mines of Plumas County, California

Unavailable since 1918, this publication was originally compiled by the California State Mining Bureau and includes details on the gold mines of Plumas County, California. Also included are insights into the mineralization and other characteristics of this important mining region, as well as the location of historic gold mines. **8.5" X 11", 200 ppgs. Retail Price: $19.99**

Mines of El Dorado, Placer, Sacramento and Yuba Counties, California

Originally published in 1917, this important publication on California Mining has not been available for nearly a century. This publication includes important information on the early gold mines of El Dorado County, Placer County, Sacramento County and Yuba County, which were some of the first gold fields mined by the Forty-Niners during the California Gold Rush. Included are not only details on early mining methods in the area, production statistics and geological information, but also the location of the early gold mines that helped make California "The Golden State". Also included are insights into the early mining of chrome, copper and other minerals in this important mining area. **8.5" X 11", 204 ppgs. Retail Price: $19.99**

Mines of Los Angeles, Orange and Riverside Counties, California

Originally published in 1917, this important publication on California Mining has not been available for nearly a century. This publication includes important information on the early gold mines of Los Angeles County, Orange County and Riverside County, which were some of the first gold fields mined in California by early Spanish and Mexican miners before the 49ers came on the scene. Included are not only details on early mining methods in the area, production statistics and geological information, but also the location of the early gold mines that helped make California "The Golden State". **8.5" X 11", 146 ppgs. Retail Price: $12.99**

Mines of San Bernadino and Tulare Counties, California

Originally published in 1917, this important publication on California Mining has not been available for nearly a century. This publication includes important information on the early gold mines of San Bernadino and Tulare County, which were some of the first gold fields mined in California by early Spanish and Mexican miners before the 49ers came on the scene. Included are not only details on early mining methods in the area, production statistics and geological information, but also the location of the early gold mines that helped make California "The Golden State". Also included are details on the mining of other minerals such as copper, iron, lead, zinc, manganese, tungsten, vanadium, asbestos, barite, borax, cement, clay, dolomite, fluospar, gem stones, graphite, marble, salines, petroleum, stronium, talc and others. **8.5" X 11", 200 ppgs. Retail Price: $19.99**

Chromite Mining in The Klamath Mountains of California and Oregon

Unavailable since 1919, this publication was originally compiled by J.S. Diller of the United States Department of Geological Survey and includes details on the chromite mines of this area of Northern California and Southern Oregon. Also included are insights into the mineralization and other characteristics of this important mining region, as well as the location of historic mines. Also included are insights into chromite mining in Eastern Oregon and Montana. **8.5" X 11", 98 ppgs. Retail Price: $9.99**

Mines and Mining in Amador, Calaveras and Tuolumne Counties, California

Unavailable since 1915, this publication was originally compiled by William Tucker and includes details on the mines and mineral resources of this important California mining area. Included are details on the geology, history and important gold mines of the region, as well as insights into other local mineral resources such as asbestos, clay, copper, talc, limestone and others. Also included are insights into the mineralization and other characteristics of this important portion of California's Mother Lode mining region. **8.5" X 11", 198 ppgs. Retail Price: $14.99**

Alaska Mining Books

Ore Deposits of the Willow Creek Mining District, Alaska

Unavailable since 1954, this hard to find publication includes valuable insights into the Willow Creek Mining District near Hatcher Pass in Alaska. The publication includes insights into the history, geology and locations of the well known mines in the area, including the Gold Cord, Independence, Fern, Mabel, Lonesome, Snowbird, Schroff-O'Neil, High Grade, Marion Twin, Thorpe, Webfoot, Kelly-Willow, Lane, Holland and others. **8.5" X 11", 96 ppgs. Retail Price: $9.99**

More Mining Books

Prospecting and Developing A Small Mine

Topics covered include the classification of varying ores, how to take a proper ore sample, the proper reduction of ore samples, alluvial sampling, how to understand geology as it is applied to prospecting and mining, prospecting procedures, methods of ore treatment, the application of drilling and blasting in a small mine and other topics that the small scale miner will find of benefit. **8.5" X 11", 112 ppgs, Retail Price: $11.99**

Timbering For Small Underground Mines

Topics covered include the selection of caps and posts, the treatment of mine timbers, how to install mine timbers, repairing damaged timbers, use of drift supports, headboards, squeeze sets, ore chute construction, mine cribbing, square set timbering methods, the use of steel and concrete sets and other topics that the small underground miner will find of benefit. This volume also includes twenty eight illustrations depicting the proper construction of mine timbering and support systems that greatly enhance the practical usability of the information contained in this small book. **8.5" X 11", 88 ppgs. Retail Price: $10.99**

Timbering and Mining

A classic mining publication on Hard Rock Mining by W.H. Storms. Unavailable since 1909, this rare publication provides an in depth look at American methods of underground mine timbering and mining methods. Topics include the selection and preservation of mine timbers, drifting and drift sets, driving in running ground, structural steel in mine workings, timbering drifts in gravel mines, timbering methods for driving shafts, positioning drill holes in shafts, timbering stations at shafts, drainage, mining large ore bodies by means of open cuts or by the "Glory Hole" system, stoping out ore in flat or low lying veins, use of the "Caving System", stoping in swelling ground, how to stope out large ore bodies, Square Set timbering on the Comstock and its modifications by California miners, the construction of ore chutes, stoping ore bodies by use of the "Block System", how to work dangerous ground, information on the "Delprat System" of stoping without mine timbers, construction and use of headframes and much more. This volume provides a reference into not only practical methods of mining and timbering that may be employed in narrow vein mining by small miners today, but also rare insights into how mines were being worked at the turn of the 19th Century. **8.5" X 11", 288 ppgs. Retail Price: $24.99**

A Study of Ore Deposits For The Practical Miner

Mining historian Kerby Jackson introduces us to a classic mining publication on ore deposits by J.P. Wallace. First published in 1908, it has been unavailable for over a century. Included are important insights into the properties of minerals and their identification, on the occurrence and origin of gold, on gold alloys, insights into gold bearing sulfides such as pyrites and arsenopyrites, on gold bearing vanadium, gold and silver tellurides, lead and mercury tellurides, on silver ores, platinum and iridium, mercury ores, copper ores, lead ores, zinc ores, iron ores, chromium ores, manganese ores, nickel ores, tin ores, tungsten ores and others. Also included are facts regarding rock forming minerals, their composition and occurrences, on igneous, sedimentary, metamorphic and intrusive rocks, as well as how they are geologically disturbed by dikes, flows and faults, as well as the effects of these geologic actions and why they are important to the miner. Written specifically with the common miner and prospector in mind, the book will help to unlock the earth's hidden wealth for you and is written in a simple and concise language that anyone can understand. **8.5" X 11", 366 ppgs. Retail Price: $24.99**

www.ingramcontent.com/pod-product-compliance
Lightning Source LLC
Chambersburg PA
CBHW081143170526
45165CB00008B/2781